D0850603

FEASTING, FOWLING
AND
FEATHERS

FEASTING, FOWLING AND FEATHERS

A history of the exploitation of wild birds

MICHAEL SHRUBB

T & AD POYSER
London

Published 2013 by T & AD Poyser,
an imprint of Bloomsbury Publishing Plc, 50 Bedford Square, London WC1B 3DP.

www.bloomsbury.com
www.bloomsburywildlife.com

Picture credits for the colour section are as follows (t = top, m = middle, b = bottom):
plate 1: Trustees of the British Museum (t, b); plate 2: Theo Daatselaar, Antiquairs (t), Imagebroker/FLPA
(b); plate 3: Cyril Ruoso / Biosphoto / SteveBloom.com (t, m); Shutterstock/Porojnicu Stelian (b); plate
4: Forschungsbibliothek Gotha (t); plate 5: Tony Marr (t); Shutterstock/David Thyberg (bl); Shutterstock/
Grant Glendinning (br); plate 6: Forschungsbibliothek Gotha (t); Shutterstock/RazvanZinica (b); plate 7:
Thomas Ross Collections (t); Shutterstock/Carmine Arienzo (m); Roger Tidman/FLPA (b); plate 8: P. A.
Morris (t); Shutterstock/Tobie Oosthuizen (bl); Shutterstock/Wolfgang Kruck (br).

ISBN (print) 978-1-4081-5990-3
ISBN (epub) 978-1-4081-6006-0

A CIP catalogue record for this book is available from the British Library

This book is produced using paper that is made from wood grown in managed sustainable forests.
It is natural, renewable and recyclable. The logging and manufacturing processes conform to the
environmental regulations of the country of origin.

Commissioning Editor: Jim Martin

Design by Julie Dando at Fluke Art
Illustrations by Alan Harris

Printed in China by C&C Offset Printing Co Ltd.

10 9 8 7 6 5 4 3 2 1

Front cover: a hunter stealthily approaches a trip of migrant Dotterel *Charadrius morinellus*;
spine: Black-tailed Godwit *Limosa limosa*, once a common breeding bird in East Anglia;
back cover: captive male Ruff *Philomachus pugnax* in the farmyard. All cover art by Alan Harris.

Visit bloomsburywildlife.com to find out more about our authors and their books

Contents

For Veronica, who was determined that
I should finish this but never lived to see it

List of Figures

CHAPTER 6

CHAPTER 7

CHAPTER 8

CHAPTER 9

CHAPTER 10

CHAPTER 11

List of Tables

CHAPTER 8

CHAPTER 9

CHAPTER 10

CHAPTER 11

CHAPTER 12

Introduction and acknowledgements

There is an extensive literature about the persecution of birds as vermin or competitors to human interests, most recently Roger Lovegrove's excellent account of the vermin records in the Churchwardens' accounts in English and Welsh parishes (Lovegrove 2007). This book deals with man's exploitation of birds for use as food, for feathers, as pets and as trophies, a field that covers both the birds themselves and their eggs.

Such exploitation has taken two forms, the domestication and breeding of species such as chickens, ducks, geese and other birds for which we use the generic term poultry, and the hunting of wild birds. A grey area lies between the two, with the large-scale rearing of gamebirds to release for shooting. In terms of food supply, much the most significant of these forms of exploitation has been domestication, but this book concerns itself with the exploitation of wild birds.

People have always hunted birds, mainly for food. Clark (1948) noted, however, that birds have rarely 'played a part in the food quest at all comparable with that of hunting (animals) or fishing'. Nevertheless, birds provided diversity in diet, which might have been particularly important in winter in mediaeval and early modern times, and often abundant seasonal food, whilst for many island communities seabirds formed an essential element of the food supply. The ingenuity expended in devising ways of catching birds and, later, the regulation of seasons and prices, argues that fowling (the taking of birds by various means) became an economically important activity.

Archaeological research in kitchen middens in prehistoric sites in Britain, Ireland, France, Germany, Denmark, Norway and Switzerland has found that important groups of species taken were wildfowl, grouse and other gamebirds, cranes and seabirds (e.g. Gurney 1921, Clark 1948, Fisher 1966, Greenway 1967, Yalden & Albarella 2009). The records came from coastal habitats in Britain and Ireland, France, Denmark and Norway, wetlands (lake villages and bog habitats) from Britain and Ireland, Germany, Denmark and Switzerland and cave sites in Britain, France and Germany, and the lists derived perhaps inevitably reflect the nature of these sites. They may also be biased by variations in the durability of the bones that form the basis of the identifications. Differing techniques of archaeological excavation also affect the range of bird species recorded. In particular most small birds will be overlooked if deposits are not extensively and thoroughly sieved (Parker 1988).

Nevertheless these groups of birds had obvious attractions for ancient fowlers. Individually, most would provide a meal. They are also gregarious, giving multiple chances to trap or to snare, and were perhaps especially vulnerable to simple fowling techniques in the breeding season or in moult. Geese and ducks, for example, could be run down when flightless, a technique used into the 20th century (see Chapter 5), and young seabirds could be taken just before leaving the nest colonies, as they still are. Yalden & Albarella (2009) noted that waders, which were favoured as food from Roman times, particularly plovers, Woodcock *Scolopax rusticola* and Curlew *Numenius arquata*, are not well represented in the prehistoric archaeological record for Britain.

More indirect evidence of the range of birds hunted by prehistoric fowlers comes from Palaeolithic and Neolithic cave paintings in Iberia, where cranes, storks, Great Bustards *Otis tarda*, wild ducks and geese, Spoonbills *Platalea leucorodia*, flamingos, Purple Gallinules *Porphyrio porphyrio*, Glossy Ibis *Plegadis falcinellus*, eagles and Marsh Harriers *Circus aeruginosus* are represented (Verner 1914). Whilst these depictions do not prove that the birds were hunted, this seems the most likely reason for depicting them. They are associated with pictures of other animals being hunted, and most of these bird species were hunted in historic times.

Ancient Egyptian art often shows scenes of wildfowling, and the birds are often clearly identifiable. Wildfowlers used S-shaped throwing sticks, weighted at one end, to break the necks of ducks in flight, and a painting from the tomb of Neb-Amon at Thebes shows him fowling, with his trained cat retrieving a passerine in its paws and a duck in its mouth (see Plate 1). Clap-nets were also used, particularly for wildfowl but also for species up to the size of cranes, and drag-nets were used for Quail *Coturnix coturnix* (Houlihan & Goodman 1986).

Many simple trapping devices have a very long history. Hobusch (1980) noted that excavations showed that many methods used in prehistoric times were still current in many parts of the world. Clark (1948) made a similar point noting that 'prehistoric man must in the main have relied upon the various types of snare and trap which occur over extensive tracts of Eurasia and North America and which still survive in parts of Europe'.

Macpherson (1897) described fowling methods from most European countries, Russia, the Middle East, Siberia, China, Japan, India, Burma, many of the Indonesian Islands, Borneo, New Guinea, Hawaii and other Pacific Islands, Australasia, both South and North America (particularly the Arctic regions), and parts of Africa. Snares, traps, nets and bird lime form the basis of the methods used everywhere. Only comparatively recently has the gun been efficient enough to supplant them, at least in more developed countries.

The historical record from the mediaeval period shows that a very wide range of bird species was taken. Macpherson (1897) provided details of fowling methods for 384 species from 67 families of birds worldwide, ranging in size from cranes and bustards to warblers and Goldcrests, taken for food, feathers, as pets or for falconry or related uses (owls, for example, were used as decoys to entice songbirds and raptors within range). Markets emerged in most major cities, and the Poulters Company regulated prices in London from the 13th century dealing, as Bourne (2003) noted, with virtually all the more edible species from southern England except scavengers, which were regarded as unclean, and raptors, although Sparrowhawks *Accipiter nisus* were eaten on Heligoland (Gatke 1895) and Scops Owls *Otus scops* on Malta (Wright 1864). Hope (1990) noted that almost no bird species found around London was deemed inedible and that birds, however small, were a valuable year-round source of fresh meat, a point also stressed by Thirsk (2007). A wide range of species we would not consider to be game today were still offered by game dealers in the late 19th century. Patterson (1905) listed 964 wild birds offered on one game dealer's stall in Yarmouth on December 16th 1889, which included 80 Blackbirds *Turdus merula*, 30 larks, 10 Moorhens *Gallinula chloropus*, 12 Water Rails *Rallus aquaticus* and 6 Dabchicks (Little Grebes) *Tachybaptus ruficollis*, and he remarked that hundreds of Blackbirds and Song Thrushes *Turdus philomelos* could be seen on other stalls the same day. Even later, Gladstone (1943) noted that, in the war years of the early 1940s, birds such as gulls, Rooks *Corvus frugilegus* and Jackdaws *Corvus monedula* were offered for sale by game dealers in London and other major cities after the availability of game ended with the shooting season. He also quoted a correspondent to the *Times* newspaper of April 14th 1942 that Red-necked Grebes *Podiceps grisegena*, Moorhens, Coots *Fulica atra* and Starlings *Sturnus vulgaris* were offered for sale in one of the great London stores, the grebes described as female geese and the Starlings as Grey Lugs!

Corporations outside London also regulated prices, in York and Hull, for example (Nelson 1907). Prices were also laid down in the household regulations and accounts of aristocratic and gentry establishments. Overseas trade developed early, and by the 19th century was worldwide in scope, a point underlined by Stubbs (1913), in a note about the availability of Asian and North American birds in Leadenhall Market in London.

Four basic questions arise in considering the exploitation of wild birds by humans – what were they taken for, how were they taken, when were they taken and what impact did fowling have on bird populations, if any. The following three chapters consider these questions in general terms. They are followed by a series of chapters on groups of species, which examine these issues in particular detail.

I have concentrated in this book mainly on the history of exploitation of wild birds in Britain and Europe, for one cannot regard Britain in isolation. A high proportion of our birds are migrants or seasonal visitors; what happens elsewhere affects their status here. Similarly, Britain has long had a significant international trade in birds, centred upon London. This was particularly true of the plumage trade that developed in the later 19th century (see Chapter 11), where London and Paris were the major entrepôts but sources were worldwide, often being colonial possessions of Britain and France.

It must also be noted that compared to the scale of exploitation and destruction that is documented in, for example, North America with the arrival and spread of European settlers (see e.g. Nicholls 2009), the historical record for Britain and Europe is relatively modest. There are perhaps two reasons for this. First, by the time that documentary records become available the land was already fully owned and controlled. Secondly, social regulation exercised some control over exploitation, either in the interests of conserving stocks, for seabirds or gamebirds for example, or by limiting those entitled to hunt. But fowling was a worldwide practice, and Macpherson's (1897) account of its history underlines three general points: the ubiquity of the pursuit of birds, the broad similarity of many of the methods used around the world, and the antiquity of many of those methods.

I am grateful to many people for help with this book. Once again I owe a considerable debt to Carole Showell, the Librarian at the British Trust for Ornithology, for her unstinting assistance in obtaining shoals of increasingly obscure references for me. She is rarely defeated and I could not operate on the outer rim of Wales without such help. The Librarians at the Public Library in Tywyn, Laura Micah and Lisa Markham, have given similar help with many books through the excellent inter-Library loan service. Their help has also been invaluable. I am also very grateful to Robert Gillmor for the extended loan of Gunnar Brusewitz's splendid history of hunting, which proved to be a most valuable reference, and to my nephew, Robert Sadler, for drawing my attention to a duck decoy at Angmering in Sussex, for obtaining a print-out of Payne-Gallwey's *Book of Decoys* for me and for prints from other old books.

A number of good friends, Roger Lovegrove, Graham Williams, Tony Marr and Martin Peers, have read through the different chapters and made many useful and valuable comments on their content and presentation, which have greatly improved the text. I also thank Alan Harris for his splendid illustrations, which so embellish the text, Jim Martin at Bloomsbury for much editorial help and advice, Elizabeth Andrews for help with German translations, and Sheila Kelly for photographing my print of partridge shooting in Plate 4.

One cannot produce a book like this without running into problems about recording currencies, values, units of area and measurement and weights, most of which have altered substantially over the centuries. For sums of money I have used the currency units used in the original sources, that is pounds (£), shillings (s.) and pennies (d.) (see Appendix 5), as I doubt if giving monetary amounts in the mediaeval period in modern pounds and pence means anything. Patently it does not if one considers the vastly different value of money today. For younger readers, note that combinations of shillings and pence are conventionally separated with a / e.g. two shillings and sixpence appears as 2/6d; a whole value for shillings may appear with a slash and a hyphen, so ten shillings is written as 10/-, while for pounds, shillings and pence the values are separated with hyphens e.g. £10-8s-6d.

From the mid-19th century I have translated sums of money into present-day values (see Appendix 5). A second problem with currencies is that Scots pounds in the mediaeval and early modern periods had a different, smaller value to their English counterparts, and documents do not always make clear which is meant. Fortunately units of area and measurement in Britain have remained constant for a long period of time, only changing with decimalisation. So giving modern values to mediaeval measurements seems to be straightforward.

Finally, I have, in general, retained the old English common names of birds, which maintains continuity with earlier sources. I have generally given the scientific name on the first appearance of each bird within each chapter. A full list of birds' names, their modern equivalents and scientific names is given in Appendix 6.

CHAPTER 1
Uses of wild birds

Food

The principal use for wild birds was and continues to be for food, and gamebirds and wildfowl remain popular items in due season today. Archaeological excavations at Roman sites in Britain summarised by Parker (1988) provided some indication of the use of birds as food in Britain in the early historic period. By far the most frequent bird remains found were domestic poultry and domesticated geese, Mallard *Anas platyrhynchos* and Pheasants *Phasianus colchicus*, indicating that rearing and eating of domestic poultry was already thoroughly established. But a wide range of wild birds was also recorded, summarised in Table 1.1. In the absence of detailed counts, the number of sites at which species were found gives some indication of how frequently they were caught.

Although the work to demonstrate unequivocally that these birds were eaten had not been done, Parker observed that one must assume that this was the chief reason for taking them, an assumption he noted as well supported by studies elsewhere, for example in Switzerland. Of the corvids, Rooks *Corvus frugilegus*, identified at 12 sites, may have been eaten, as they still are in Rook pie, but Ravens *Corvus corax* were certainly kept as pets (as they were in the Victorian era; Ratcliffe 1997), which must account for some records, and it seems very probable that this also applies to Magpies *Pica pica* and Jackdaws *Corvus monedula*. There was a strong Roman tradition of teaching these species, as pets, to talk (Toynbee 1973). Corvids probably also occurred in these records as a result of their scavenging habits.

There were strong similarities with the prehistoric period, in the importance of wildfowl and gamebirds. Waders assumed much greater importance in Roman times, which also saw the introduction of domestic pigeons and doves to Britain (Yalden & Albarella 2009). Many songbirds were eaten, particularly thrushes, a long-standing Roman tradition (Gurney 1921).

Species group	Number of species	Number of sites	Notes
Herons	2	4	
Storks		1	
Swans	3	3	
Greylag/domestic goose		29	
Other geese	5 plus unidentified	9	
Shelduck		1	
Mallard, Wigeon, Teal		53	Mallard at 41 sites
Other dabbling ducks	3	5	Pintail not recorded
Diving ducks	6	9	
Gamebirds	5 plus unidentified	21	
Rails	4*	10	
Crane		12	
Great Bustard		1	
Plovers	3	18	Golden Plover at 13 sites
Snipe and Woodcock		29	Woodcock at 27 sites
Other waders	11	16	
Pigeons and doves	4 plus unidentified	34	Domestic/Rock at 11 sites
Seabirds	5	6	
Skylark		3	
Thrushes	5 plus unidentified	20	
Other small passerines	16 including unidentified	21	Starling at 12 sites
Corvids	6	57	Raven at 39 sites

Table 1.1. *Wild birds recovered at 86 Roman sites in Britain. Data from Parker 1988. * includes Coot and Moorhen.*

Parker (1988) commented that dietary customs in Britain were substantially Romanised, and that the lack of seabirds, commonly taken in north and west Britain, reflected the south-east geographic bias of the Roman province. Yalden & Albarella (2009) summarised records from more Roman sites, but these do not materially alter the pattern shown in Table 1.1.

Excavations from mediaeval sites show this pattern continuing, with a preponderance of domestic chickens and geese from the 7th to the mid-16th centuries (Serjeantson 2006), but the proportion of wild birds taken was very variable (see also p. 21). Table 1.2 summarises the archaeological records from mediaeval sites in southern England. Serjeantson noted that the species of wild birds most frequently found in Saxon sites were Grey Partridge *Perdix perdix*, Lapwing *Vanellus vanellus*, Woodcock *Scolopax rusticola*, Eurasian Curlew *Numenius arquata*, Woodpigeon *Columba palumbus* and thrushes. Less frequent were wild duck, Golden Plover *Pluvialis apricaria*, Snipe *Gallinago gallinago* and godwits. From the mid-11th century the main species were Grey Partridge, plovers, Woodcock and Snipe. Many swans have also been found.

Recorded in 1–5 sites	Recorded in 6–10 sites	Recorded in >10 sites
divers	Cormorant	Mute Swan
grebe	Grey Heron	Grey Partridge
shearwater	Coot	plovers
Gannet	Crane	Woodcock
Shag	Lapwing	Snipe
Bittern*	Ruff	thrushes
Stork*	godwit	
Spoonbill*	Curlew	
Quail	gulls	
Great Bustard*		
Moorhen		
Dotterel		
Small waders		
Whimbrel		
tern		
Common Guillemot		
lark		
sparrow		
other small birds		

Table 1.2. *Wild birds recorded in archaeological records from mediaeval sites in southern England. Data from Serjeantson 2006. Note that there are also uncertain identifications for Water Rail, Oystercatcher, Razorbill and Puffin. * indicates one record only.*

Feasts and market lists

Other important sources of information on the birds consumed in mediaeval and early modern Britain are household accounts and lists of birds eaten at notable feasts or recorded in corporation market price lists, many recorded in the 19th and early 20th century ornithological literature. Table 1.3 summarises the frequency with which species occurred in 58 such lists, to give some idea of the range and variety of wild birds that were then eaten. Records of plovers, of small waders and of gulls are grouped together as it is not always clear which species were meant. The numbers of birds involved in these sources was not stated consistently. Nevertheless household accounts do show a similar pattern to the archaeological record, with Grey Partridges, plovers, Woodcock and Snipe as the most abundant species recorded in the accounts edited by Woolgar (2006), for example.

A similar source of information on the birds commonly eaten at that time is provided by price lists of the Poulters Company of London, given by Jones (1965). Table 1.4 summarises 23 of these price lists, from the late 13th century to the end of the 16th century. Domestic poultry (including geese unless specifically stated as wild) has been excluded. Pigeons have

Species	Number of lists	Species	Number of lists
Bittern	27	Pheasant	28
Night Heron (Brewe)	8	Crane	25
Egret sp.	13	Great Bustard	7
Grey Heron	33	Oystercatcher	4
Stork	4	plovers	43[2]
Spoonbill	10	Snipe	26
Swan (probably Mute)	38[1]	Woodcock	33
Wild geese	9	Curlew	32
Shelduck	2	Redshank	6
Wigeon	6	Other waders	22
Teal	28	gulls	15[3]
Mallard	26	Puffin	4
Wildfowl	6	Pigeon (dovecote?)	28[4]
Grouse (three species)	10	Pigeon (wild)	8
Grey Partridge	42	Skylark	31
Quail	19	Other passerines	28
Peacock	19		

Table 1.3. *The frequency with which species of birds occurred in household accounts, menus from feasts, and in Corporation market price lists in mediaeval and early modern Britain. Domestic poultry are excluded. Figures are the number of lists out of 58 in which the species occurred. Data from Pennant 1776, Harting 1879, Yarrell & Saunders 1884, Lennard 1905, Nelson 1907, Stubbs 1910a, Le Strange 1920, Gurney 1921, Ticehurst 1923, 1934, Mead 1931, Simon 1952, Darby 1974, Rackham 1986, Bourne 1981, 1999b, 2003, 2006, Woolgar 1999, 2006. Ticehurst 1923 is treated as one list;* [1] *includes one record of Whooper Swan;* [2] *probably mainly Golden Plover but here includes Dotterel, Lapwing and Ringed Plover;* [3] *mainly Black-headed Gull (or Puets);* [4] *includes 11 squabs.*

also been excluded as dovecote birds unless otherwise stated. Swans, although probably all Mute Swans *Cygnus olor*, are included. 'Green' and 'Grey' Plovers (which probably equate to Golden Plover) and Lapwing are treated as 'plovers' since it is not always clear which species is meant. Blackbirds are included as thrushes, and finches and sparrows are combined.

Several points must be made about these lists. First, those in Table 1.3 are drawn mainly from feasts or household books of aristocratic and gentry establishments. They are thus possibly biased towards what was consumed by the upper classes, and should not be taken as representative of the diets of ordinary people. Nevertheless Thirsk (2007) pointed out that cookery books from the 16th century at least show much interest in bird meat, and recipes were included for many of the species listed in the table. It seems clear that by then, if not before, such food was eaten by a wider sweep of society than just the upper classes. Secondly, in the aftermath of the Black Death, the number and range of birds, including passerines, eaten regularly rose considerably. Stone (2006) gave the example of

Species	Number of lists	Species	Number of lists
Bittern	20	Dotterel	1
Night Heron (Brewe)	6	plover	22
Egret sp.	6	Knot and Ruff	1 and 1
Grey Heron	20	Snipe	22
Stork	4	Woodcock	23
Spoonbill	14	godwit	1
Swan (Mute?)	21	Curlew	14
Wild goose	1	Redshank	1
Wigeon	4	gull	2
Teal	22	mew (probably a gull)	1
Mallard	23	Stock Dove	2
Grey Partridge	19	larks	19
Quail	4	thushes	9
Pheasant	19	finches	5
Peacock	2	bunting	1
Crane	15	Great birds	6
Great Bustard	9	Little birds	5

Table 1.4. *Bird species listed in price lists of the Poulters Company of the City of London over the course of three centuries. Figures are the number of lists out of 23 in which the species occurred. Data from Jones 1965.*

London cook shops which were then offering birds such as roast heron, capon pasties, roast Woodcock, thrushes, larks and finches – what he termed high-class fast food. Swans and herons were eaten more frequently, and estates increasingly diversified into managing semi-wild birds, developing swanneries in the 14th and 15th centuries, for example. Stone also suggested that heronries were similarly managed (see p. 58). These patterns were part of a trend toward the diversification of agriculture and food supply after the Black Death, as cultivation became more difficult with a declining population, and cereals became less profitable (see also Thirsk 1997).

Thirdly, except for Common Cranes *Grus grus*, various herons and swans, the lists continue to be dominated by the familiar staples we still find in field sports today – gamebirds, ducks and geese, and waders. Although fewer wader species are taken today, those listed were common quarry well into the 20th century. Similarly, although songbirds are no longer taken for the table in Britain, many were taken into the early 20th century (see p. 14). Skylarks *Alauda arvensis*, for example, were still offered by poulterers in London in the early 1940s (Gladstone 1943), and cookery books of the same era also offered recipes for preparing them for the table. It was also the fate of many female songbirds caught by bird trappers engaged in the cagebird trade up to the end of the 19th century to be killed and sold off for food. Songbirds remain part of the diet in southern Europe.

Conspicuous consumption

Astonishing numbers of Bitterns *Botaurus stellaris*, herons, Spoonbills *Platalea leucorodia*, storks, swans and cranes were sometimes consumed at feasts. The great feast in September 1465 at the installation of George Neville as Archbishop of York is a case in point, which has been widely quoted. The list of provisions, as transcribed by Pennant (1776) from Leland's *Antiquarii Collecteana,* is shown in Table 1.5.

Further particulars of the actual courses also listed Redshanks *Tringa totanus*, stints, larks and 'Martynettes roast'. Pennant believed the last to be Swallows *Hirundo rustica*, which Macpherson (1897) noted were taken in large numbers in northern Italy, being much in demand by the poulterers of Padua, for example. Both Pennant (1776) and Woolgar (1999) indicated that the oxen, sheep, pigs and what Pennant termed 'other more substantial foods' were served to the retinues of the noblemen present, amounting to more than 400 persons.

Item	Number	Item	Number
In wheat	300 quarters	Pigeons	4000
In ale	300 tuns	Rabbits	4000
Wine	100 tuns	In Bitterns	204
Of hippocras	1 pipe	herons	400
In oxen	104	Pheasants	200
Wild bulls	6	Grey Partridges	500
Muttons	1000	Woodcock	400
Veales (calves)	304	Curlew	100
Porkes	304	Egrittes	1000
Swans	400	Stags, buck and roe	500 and mo.[re?]
Geese	2000	Cold venison pasties	4000
Capons	1000	Parted dishes of jellies	1000
Pygges	2000	Plain dishes of jellies	3000
Plovers	400	Cold baked tarts	4000
Quail	100 dozen	Cold baked custards	3000
The fowls called Rees	200 dozen	Hot venison pasties	1500
In peacocks	104	Hot custards	2000
Mallard & Teal	4000	Pike and bream	608
In Common Cranes	204	Porpoises and seals	12
In kids	204	Spices and sweetmeats	plenty
In chickens	2000		

Table 1.5. *Provisions for the feast celebrating the installation of George Neville as Archbishop of York in September 1465, from Pennant 1776. For 'a quarter of wheat' see Appendix 5. Hippocras was an aromatic medicated wine much used as a cordial. No explanation is given for the appearance of both porkes and pygges; perhaps pork and bacon? No explanation is given for the term 'In cranes', 'In chickens', etc. 'Rees' are usually interpreted as Ruffs but see p. 113.*

However, both Gurney (1921) and Bourne (1999) were sceptical that all these provisions were actually consumed. But it was not expected that those present at such functions would eat everything. A substantial amount of broken meats, or left-overs, was expected to be given as alms to the poor, and aristocratic establishments at this time employed almoners, whose duties included responsibility for overseeing the gathering and distributing of such post-feast gifts (Woolgar 1999).

Nor was such scale at a major feast unique. Hobusch (1980) listed, from State Papers, the game consumed at the wedding feast of the Polish Duke Johann Sigismund in 1594, which comprised 13 Bison, 20 Elks, 10 Red Deer, 22 does, 36 Wild Boar, 29 sounders (young pigs), 2 bears, 48 Roedeer, 272 hares, 5 wild swans, 123 Woodcock, 279 heath cocks (Black Grouse *Tetrao tetrix*), 433 Hazel Grouse (*Bonasa bonasia*), 47 partridges and 413 wild ducks. The provisions requisitioned by Henry III for his Christmas feast in 1251 involved 430 Red Deer, 200 Fallow Deer, 200 Roedeer, 200 wild swine, 1,300 hares, 450 Rabbits, 2,100 partridges, 290 Pheasants, 395 swans, 115 Common Cranes, 400 tame pigs, 70 pork brawns, 7,000 hens, 120 peafowl, 80 salmon and lampreys without number (Rackham 1986), and for a feast for King Richard II in 1387 50 swans, 200 geese, 120 Curlews, 144 Night Herons *Nycticorax nycticorax* and 12 Common Cranes were among the viands provided (Stubbs 1910a). Bourne (1981, 1999b) also listed birds consumed at the Field of the Cloth of Gold in 1520, and at the meeting between Henry VIII and the King of France at Calais in 1532, which included totals of 86 Bitterns, 801 Brewes (Night Herons), 440+ Grey Herons *Ardea cinerea*, 304 Common Cranes and storks, 65 Spoonbills, 48 Great Bustards *Otis tarda*, 361 swans, 1,800 partridges, 5,947 Quail *Coturnix coturnix* and 3,120 Snipe; the list for 1520 included a separate item of 912 Bitterns, Curlews, shovellers (Spoonbills) and gulls.

There is clearly a strong element of conspicuous consumption involved in these lists. Eating species such as herons, cranes and swans was an important mark of status. The birds listed for the Neville feast would, on the basis of prices given for around the same period by Jones (1965), have involved an expense of *c*.£260, a very large sum of money in the mid-15th century, when a labourer's daily wage was around 2–3d. The cost of the birds consumed in 1520 and 1532 was even greater and Bourne (1981) estimated it at £545 and eight shillings. These lists are also markedly at variance with the patterns shown by household accounts, which more usually record the acquisition and consumption of such birds in ones and twos, occasionally tens and twenties, when they would have been penned and kept for future consumption.

There were elaborate rules for dressing and carving birds such as herons and cranes. An early 16th century 'Boke on Kervynge', quoted by Stubbs (1910a), noted that one displayed a Crane, dismembered a Heron, unjointed a Bittern, broke an Egret, and minced a Plover (modern spellings). Whilst one may now not understand what these terms meant, they indicate that prescribed ceremonial methods of carving such birds were followed. Hope (1990) also noted that swans and peacocks were commonly skinned, rather than plucked, and roasted and then reassembled to be served in all their feathered beauty, with bills and some feathers further decorated by gilding. Gurney (1921) gave a recipe for Peacock *Pavo cristatus* in which the bird was 'flayed, parboiled, larded and stuck thick with cloves; then roasted, with his feet wrapped up to keep them from scorching; then covered again with his own skin as soon as he is cold, and so underpropped that, as alive, he seems to stand on his legs'. Such consumption and ceremony contributed to the projection of the power, wealth and prestige of the giver of the feast.

Figure 1.1. *A 17th-century street market. The stall on the left offers a peacock and, to its right, what is probably a swan, a crane (which the stall-holder is lifting down), a large wader (perhaps a Curlew) and a goose. From Burke (1940).*

Although egrets, Night Herons, Spoonbills, storks and Common Cranes were apparently no longer offered for sale in England much beyond the end of the 16th century (see Figure 1.1), Bitterns, Grey Herons and Great Bustards continued to appear in the markets into the early 19th century (Gladstone 1943). Spoonbills and Common Cranes ceased breeding in Britain in the early 17th century, although Common Cranes remained winter visitors, and imported species, such as Night Herons, may have become more difficult to obtain, becoming increasingly scarce due to over-exploitation (see Bourne 1999b).

There was a considerable import trade in live birds through Calais by the early 16th century, which has been well summarised by Bourne (1999b), reviewing information in the Lisle letters from Calais relating to the English bird trade. This trade mainly concerned birds intended for food or for falconry. The main species imported for food from mainland Europe at this time were herons – Grey Herons, Night Herons and egrets – and Quail, with lesser numbers of larger gamebirds, Mute Swans, Common Cranes, storks, and occasionally songbirds (to be kept as pets). Lapwing, Snipe and Dotterel *Charadrius morinellus* were also sent, although many of the latter were by then being obtained from Lincolnshire. Anne Boleyn was fond of Dotterels, and kept them in her garden until it was time to eat them. Lisle specialised in Quail, which he bought in vast quantities as presents (see Chapter 6). Such trade was international. Thirsk (2007), for example, recorded fattened birds being exported from Poland to the Netherlands in the 17th century, sent live in baskets in the corn boats from Danzig.

Smaller numbers of gulls and preserved Puffins *Fratercula arctica* (counted as fish by the Church and therefore permissible on Fridays and in Lent) were sent from England to France. Peregrines were also sent from England, whilst Goshawks, particularly, came from France. The trade continued after Calais reverted to France in the mid-16th century, and William Harrison, writing in 1577, noted the continued import into England of 'egrets, pawpers (Spoonbills) and such like … daily brought unto us from beyond the sea'. These birds were shipped live, the herons and gulls mainly taken as young from the nest and sent to be reared in pens or houses, known as stews or mews, to be eaten when required. Methods of preparation described by Stubbs (1910a) show quite clearly that these birds were killed and dressed as required for the table. The costs of making pens to hold them, of food for the birds and of keepers to look after them appear not infrequently in household accounts of the period. Quails were sent in baskets provided with hempseed and water for the journey (Hope 1990).

Stews or mews

The art of capturing birds alive and fattening them in stews or mews was practiced with species besides herons and gulls. It was commonly done with waders (see Chapter 7), with gamebirds and Turtle Doves *Streptopelia turtur*, and with thrushes and Ortolan Buntings *Emberiza hortulana* (see Chapter 9). This is an ancient practice and it was frequently illustrated in ancient Egyptian art, particularly featuring cranes, herons and geese. Cranes are shown being herded by keepers, who guided them with long sticks. Both cranes and geese were crammed *i.e.* force-fed. Foods such as hempseed, milk curds, wheat, barley and liver were widely used, and wildfowl were coaxed to feed on wine and ale, which made them drunk. Birds were often placed with others already tamed for reassurance (Thirsk 2007). Young Black-headed Gulls *Larus ridibundus*, known as puets, and herons were fed on ox liver and household scraps and sometimes curds to fatten them and to sweeten the flesh. Herons at Althorp were also fed on oatmeal (Lilford 1895). Even Spoonbills were recorded as readily taking such foods. Thirsk also noted a fashion in the early 17th century in parts of eastern England for breeding species such as Quail and plovers for meat and eggs.

Being outside the restrictions of the Game Laws in England (see Chapter 5), a large trade developed in wildfowl and waders, with wildfowl particularly important following the introduction and spread of Dutch-type duck decoys from the early 17th century. Virtually all waders were taken for the table, but the records show a preponderance of plovers, Snipe, Woodcock and Curlew (see Chapter 7).

Passerines were eaten by all classes of society, a tradition that went back to classical times at least. Although the point should not be exaggerated, Macpherson (1897) remarked that the English have never indulged in the destruction of small birds for the table to the same extent as the French, the Germans or the Italians. It is probable that the restrictions placed by the scale of enclosure and game preservation from the late 18th century on the activities of bird-catchers in England partly underlie this observation (see Chapter 9). Enclosure in France, Germany and Italy was a far more piecemeal and fragmented process (Blanning 2007).

In his account of the history of fowling Macpherson (1897) makes it clear that all groups of seabirds were exploited, mainly for food, worldwide (see Chapter 8). Seabird colonies are an obvious target for such exploitation combining, as they do, concentrations of birds (and eggs) of good size and abundance in one place. Thus Gannets *Morus bassana* and auks, and, latterly, Fulmars *Fulmarus glacialis*, for example, were widely taken for food by local communities at stations all round the North Atlantic from prehistoric times.

Changes in attitudes

With the development of efficient sporting firearms and increasingly efficient and abundant food production from the mid-18th century, attitudes to taking birds in Britain changed. Wild birds were no longer regarded as part of the produce of an estate to be harvested to help feed the household. Instead, gamebirds and wildfowl came to be valued for field sports and the management of estates at least partly devoted to this purpose. 'Vermin' was rigorously controlled, methods of rearing and feeding game were developed or expanded and coverts laid out to improve the sporting nature of the shooting. It is probably no accident that the age of the big shots (see Ruffer 1977) coincided with a period of marked agricultural depression in the late Victorian and Edwardian eras. Estates had often developed or acquired other sources of income and, as agricultural prosperity declined, landowners increasingly regarded the sporting aspects of their estates more highly. Tenants were, for example, prevented from reorganising the field patterns of their holdings to preserve hedges for game and shooting, a significant check on agricultural improvement and profitability (Shrubb 2003). Markets also dealt with a narrower range of species, mainly gamebirds, wildfowl and some waders (e.g. Gladstone 1943), although passerines, particularly larks, were still traded into the 20th century.

Feathers and down

Feathers are an inevitable by-product of eating birds, and have had many uses, for example as quill pens, for firescreens, for stuffing pillows, cushions, quilts and mattresses, for fashionable decoration of hats and clothes and as fletching for arrows. Wildfowl were perhaps the most important group for many of these uses (see Chapter 5), although there was also an extensive trade in the feathers of seabirds (Chapter 8). Minor uses included making fishing flies (trapped Wrens *Troglodytes troglodytes* were released without their tail feathers, plucked for this purpose; Swaine 1982) and as paint brushes. Payne-Gallwey (1882) noted that the Great Northern Diver *Gavia immer* was 'well worth a little trouble, for if not in sufficiently good plumage to please a collector, the large white breast makes the perfection of a fowling cap, and three such skins an excellent waistcoat. Impenetrable to wet, tough as leather and warm, the plumage is of a most suitable kind and colour for a fowler'. Smith (1887) noted that the skins of divers were much favoured for this purpose in Scandinavia.

Geese

Domestic geese were particularly important for quill pens and soft furnishings and, in the mediaeval period, for fletching arrows. In mediaeval Britain, geese were kept in small units by peasants, but by the end of the period they were also being raised in large flocks (Gurney 1921, Serjeantson 2006). Such flocks were kept on commons, particularly the commons of the Fens and Somerset, for both meat and feathers. For meat the London markets, for example, were supplied by geese driven from the eastern counties (Figure 1.2), particularly from Norfolk and Suffolk. Defoe (in Furbank *et al.* 1991) described the goose drives from these counties, saying 'a prodigious number are brought up to London in droves from the farthest parts of Norfolk; even from the fenn-country, about Lynn, Downham, Wisbich and the Washes; as also from all the east-side of Norfolk and Suffolk, of whom 'tis very frequent now to meet droves, with a thousand, sometimes two thousand in a drove: they begin to drive them generally in August, by which time harvest is almost over, and the geese may feed on the stubbles as they go'. Driving ceased in October, when the roads became too muddy for the geese to negotiate.

There were important feather industries in Lincolnshire and Somerset in the 18th century, supplying feathers and down for stuffing mattresses. Pennant (1776) recorded that the geese in Lincolnshire in his day were plucked five times a year, from Lady-Day (March 25th), when feathers and quills were taken, and four times subsequently for feathers, until Michaelmas (September 29th). They were plucked alive and Lord Orford, in his *Voyage through the Fens*, recorded having found many dead and dying geese as a result of this

Figure 1.2. *A goose drive,* en route *to Boston from Kirton Lincolnshire in 1877. From* Rowley's Ornithological Miscellany *of 1878.*

practice (Wentworth-Day 1954). Pennant noted that the feathers 'are a considerable article of commerce; those from Somersetshire being esteemed the best; and those from Ireland the worst'. Every four birds yielded a pound of feathers, worth 9d in the 1740s. This cottage industry was progressively killed off by the enclosure of the commons from the late 18th century and the resultant loss of free grazing to the commoners. In the late 19th century, around 70% of the feathers used in Britain (775 out of 1,075 tonnes) were imported, mainly from Russia. Feathers remain a significant product in Europe; for example, goose farms with large flocks of white geese, which are still plucked alive, are a feature of the Hungarian Plain (Kear 1990). In Britain, however, feathers generally come from dead geese.

Hardy (1992) gave some detail of the demand for goose feathers for fletching arrows in mediaeval England during wartime. In February 1417, six feathers from every goose in 20 southern counties were ordered to be sent to the Tower by March 14th, and, on December 1st 1418, sheriffs were ordered to supply 1,190,000 goose feathers by Michaelmas. Such orders were issued annually to replace stocks.

Feathers as decoration

Feathers also have a long history as decorative objects. Egret plumes were prized in the East long before they became fashionable in the West, and the Polynesians prized the red and yellow feathers of certain species for feather cloaks and head-dresses, particularly those of honeycreepers in Hawaii (Greenway 1967, Diamond 2005, Hume & Walters 2012). The Maoris collected Huia *Heterolocha acutirostris* skins as status symbols and for ornamentation of the dead. They were often carefully preserved in special caskets as heirlooms (C. Simsom in Chalmers-Hunt 1976). American Indians made feather bonnets from eagle feathers and fletched their arrows with them, too, in the belief they would enhance their shooting. The impact of such exploitation was limited because these were all objects of status and prestige.

This was not the case with the fashion trade that emerged in the 19th century, when straight monetary value created worldwide and immensely damaging demand (see Chapter 11). Much of the literature on this subject deals particularly with herons and egrets. But brightly coloured species such as Bee-eaters and Rollers were similarly persecuted.

Cagebirds

A third major exploitation of wild birds was and remains the trade in cagebirds. The keeping of birds as pets has a very long history which is dealt with in detail in Chapter 10. But in Britain, at least, changes in land use, particularly enclosure and strict game preservation, together with progressively more comprehensive bird protection legislation during the 19th century, increasingly circumscribed the opportunities for trapping native songbirds for the cage, a point confirmed by Yarrell & Newton (1874).

Aviculture now increasingly rests on imports, and the cagebird trade today is largely centred on the Far East, and parts of Africa and South America. Inskipp & Gammell (1979) estimated that just under seven million birds were exported from these regions in the early 1970s, with India (38%), Senegal (28%), Indonesia (7%) and Thailand (6%) accounting for 79% of that total. More than half the birds were imported into Japan, the world's largest importer of cagebirds, the United States and western Europe. The numbers imported into the United States and western Europe declined sharply during the 1970s, with the introduction of disease control regulations that remain in force.

Consequences of trade

Inskipp & Gammell (1979) noted two important consequences of this trade. In the country of export overtrapping, perhaps combined with habitat destruction, may threaten the survival of target species, although they concluded any such effects were then unknown. In the country of import exotic species may represent a threat either by introducing diseases (less likely with the quarantine systems now in force) or by escaping and establishing feral populations, which may lead to damaging competition with native species for scarce resources such as nest sites. Parrots are a particular case, with ineradicable feral populations of Ring-necked Parakeet *Psittacula krameri* established in southeast England, centred on London, and in at least five other European countries (Lever 2005), and similar populations of Monk Parakeets *Myiopsitta monachus* established throughout the United States but particularly in Florida and California (Sibley 2000). Sibley included 27 species of parrots and allied species in his guide, none native, and observed that more than 65 species had been recorded in Florida, all of which were escapes from captivity.

Inskipp & Gammell also noted significant mortality in cagebirds between the point of capture and the point of sale, and in quarantine. Most birds were also destined to become pets and would have had no chance to breed. Both these factors created further demand, so supporting the persistance of the trade.

Falconry

Falconry originated in Asia, probably in China between 689 and 675 BC, and reached Europe in the third century AD (Hobusch 1980), coming into England with the Saxons (Gurney 1921). Falcons and hawks were valuable property kept by the upper classes for sport and to supply the larder. Because of their value for these purposes and the status their possession conferred, they were strictly protected. Certain species were used for particular forms of hunting; for example, Merlins *Falco columbarius* for larks, Peregrines *Falco peregrinus* for herons and kites, and Gyrfalcons *Falco rusticolus* for kites and cranes. The three most favoured species were Gyrfalcons, Peregrines and Goshawks, although Newton (1879) pointed out that 'goshawk' did not always refer to *Accipiter gentilis* in the past. As a crude generalisation, falcons were flown primarily for sport, while Goshawks were valued larder-fillers. Yalden & Albarella (2009) drew attention to records of hawk's eyries following entries for woodland in north Wales and eight English counties in Domesday Book, particularly Cheshire where 24 were recorded. They concluded that these entries referred to Goshawks because of the high value placed upon them (up to £10), and the population density indicated for Cheshire, which was about equivalent to that found in Britain today.

Hobusch (1980) drew attention to the thousands of falconers and their birds employed by Asian potentates in the mediaeval period. There was also a considerable trade in falcons and hawks in Europe from at least the 11th century, and perhaps as early as the 8th century (Yarrell & Saunders 1884). In England Gurney (1921) recorded that fines (a form of taxation rather than punitive) were often paid in falcons in the 12th century and later, and frequently involved the highly prized Gyrfalcons; white morphs were especially valued. Large numbers were sometimes involved. One Lincolnshire gentleman was obliged to find '100 Norway Hawks and 100 Girfals' in 1131, presumably all Gyrfalcons, and four of the Norway Hawks and six Girfals were specified to be white. Most of these birds were probably imported, and falcons were regularly brought into England from Scandinavia, Russia and France by the nobility. But some were taken as birds on passage within the British Isles. James VI/I certainly obtained such birds from Ireland (then part of the British Crown), paying £70 in 1624 for 'three gierfalcons and a jerkin [male Gyr], bought by the King from the hawk-taker of Ireland' (Macgregor 1989), perhaps providing evidence that the species was a more regular winter visitor in the early modern period than it is today. There are records of this trade going back to the 14th century (Woolgar 1999).

Harting (1890) also noted that many lands in Essex and other counties in England were held in the mediaeval period by serjeanty (a form of feudal tenure on condition of service to the King only) of keeping hawks and finding hounds for the King's use, should he require them. The birds involved included Lanners *Falco biarmicus*, which must have been imported, kept for heron hawking.

Passage hawks or adults taken from the nest were preferred to eyasses (young taken from the nest to be reared and trained), as they had already been trained to hunt by their parents. James VI/I expended large sums of money on falconry. For example, he paid his Master of Hawks £243–4s for hawks in May 1618. Costs also included food for the hawks, for example payment of £24–6s–8d for 30 dozen pigeons in 1608, and payment of £131–7s–6d for feeding and keeping herons at Theobalds (a royal estate in Hertfordshire), presumably to enter and train heron-hawks (Harting 1880). Payments of this nature, although not of this scale, feature regularly in mediaeval household accounts, with the acquisition of hawks, of chickens, pigeon squabs and meat to feed them, and of materials for manning them being accounted for.

Appendix III in Pennant (1776) also shows that James VI/I expected to have first choice of any hawks and falcons that were imported for sale. The warrant, dated January 26th 1621, stated

> *To all those to whom this present Writing shall come I, Sr. Anthony Pell Knight Master Falconer Surveyor and Keeper of his Majesty's Hawks send greetings. Whereas I am credibly informed that divers persons who do usually bring Hawks to sell do commonly convey them from shipboard and custom house before such time as I or my servants or deputies have any sight or choice of them for his Majesty's use whereby his Highness is not nor has not lately been furnished with the number of hawks as is most meet.*

The warrant went on to command all those who imported hawks not to remove them from ship or custom house until the King's representatives had had their choice for the King's service. Prices were fixed at 26/8d for female Peregrines and Lanners, 20/- for female Goshawks and 30/- for female Gyrs, and at 13/4d for males of these species. Whether these were the going rates in the open market is not stated.

Trapping raptors

Hawks on passage were frequently taken by bird catchers netting flocks of migrating birds and taken to courts or noblemen, who paid standard prices for them. Trapping hawks was a profitable occupation in some areas, for example the Crau near the Carmargue in southern France, where the best Lanners were trapped, and on the plains of Brabant. The plover-netters on the Lincolnshire Wash and Humber also regularly trapped Peregrines and Merlins in the 19th century, and some of these, at least, were sold on to falconers (Lorand & Atkin 1989). But the finest birds of prey came from Scandinavia and further northeast in Europe. Many European courts were supplied from the training school for falconry, founded in 1390 by the Order of Teutonic Knights at Marienburg (now Malbork in northern Poland). Konigsberg (now Kaliningrad), also a city of the Teutonic Knights, was another important source, sending birds taken in the Baltic area over much of central and western Europe. Demand was constant, and a lucrative trade developed throughout Scandinavia, western Europe, the Balkans, the Middle East and North Africa. An important staging post for this trade in the 14th century was Bruges, from where merchants dispersed the birds throughout Europe. King's Lynn in Norfolk was a regular entry port for these birds in England, and falcons were sometimes shipped there from Norway in direct exchange for East Anglian grain. Hungary also exported falcons, most probably Sakers *Falco cherrug*, which were also taken in Romania, and the Knights of St John on Malta were another major source. Danish falconers caught and imported 'shiploads' of white Gyrfalcons from Greenland, which earned the Greenlanders large sums of money into the 18th century. One ship, for example, brought 148 Gyrfalcons from Iceland to Copenhagen in 1754 (Lloyd's *Scandinavian Adventures* in Salvin & Brodrick, 1855). Hobusch (1980) showed a painting of the falconer of one German Margrave carrying a cadge of 12 white Gyrfalcons in 1752. Such cadges – padded square frames upon which the falcons sat, suspended from the falconer's shoulders (see Figure 4.3) – were the standard way of transporting them (Hobusch 1980, Cummins 1988).

Besides such trade, falcons were important diplomatic gifts between rulers. Hobusch (1980) noted, for example, that the Grand Master of the Teutonic Order presented 1,118 falcons to various foreign princes and rulers between 1533 and 1569. It is not altogether surprising that the stock of desirable falcons declined sufficiently rapidly in the late Middle Ages that the demand for young birds could scarcely be satisfied. Such diplomatic presents could also be expensive for the recipient. James VI/I once disbursed a total of £739–7s–4d to provide French gentlemen bringing him hawks from the King of France with entertainment spanning 35 days (Harting 1890).

Valkenswaard

The centre of the falcon trade in the 18th century moved to Valkenswaard in the middle of the extensive area of heath, moor and fen known as the Kempen, which straddled what is now the Dutch/Belgian border south of Eindhoven. From October 1st, falconers there erected simple turf huts from which they operated an intricate system of poles, lines, decoys and bownets to catch Peregrines. Great Grey Shrikes were placed on nearby knolls to warn of the approach of a falcon. The catching season lasted for three months, and the long process of manning and training the birds started as soon as they were caught. In the early spring the falconers returned to their employers, among which the royal courts of England, France, Germany and Portugal were prominent.

Falconers from elsewhere also came to Valkenswaard to buy falcons, for the hawk-catchers of the region also undertook expeditions to Iceland, Norway, Sweden and western Russia as far north as Murmansk to catch birds, particularly Gyrfalcons. These they took back to trade at Valkenswaard. Licences for this trade were issued by the Danish kings, and the Valkenswaard men had a monopoly in catching and trading. Between 1731 and 1793, 4,649 Gyrfalcons were imported (van de Wall 2004).

Interest in falconry declined sharply from the late 18th century as the sport of shooting flying birds developed, with the production of more efficient guns (though there remains to this day an international trade in falcons, often illicit, which particularly involves supplying Gyrs, Sakers and Peregrines to Arab potentates). The development of intensive game preservation from the early 19th century saw the status of all raptors, in northwest Europe especially, change from birds valued and protected for sport to pests that needed to be eliminated. The scale of this persecution was well described by Bijleveld (1974). In his preface the author remarked that the present scarcity of birds of prey in Europe represented an abnormal situation. In the absence of any systematic information about birds of prey in the mediaeval and early modern periods, the impact of falconry on wild populations cannot be assessed, but it was likely to have been significant.

Cormorants

King James VI/I also kept Cormorants *Phalacrocorax carbo*, and was accustomed to travelling about the country with them, fishing as he went (Harting 1871). This ancient technique probably originated with the Chinese, who still practise it today (Figure 1.3). King James had a regular establishment for his Cormorants on the Thames at Westminster, and created the office of Master of the Royal Cormorants for their keeper, but this office did not survive the Civil War of 1642–49 (Harting 1871). Harting noted that this official's full title described

Figure 1.3. *Fishing with Cormorants in China. From contemporary descriptions, the method used in England in the 16th and 17th centuries was similar. From an engraving in Yarrell & Saunders (1885).*

him as keeper of the His Majesty's Cormorants, Ospreys and Otters, which implied that attempts were made to train the latter two species for fishing, presumably without success. Fishing with Cormorants was still practiced in England up to the end of the 17th century (Ray 1678) but there seems to be no record of it thereafter.

Eggs and egging

Wild birds did not only come to table as a result of hunting. The taking of the eggs of wildfowl and gamebirds and hatching and rearing the young for food was well understood by the 16th century, and almost certainly long before. Thirsk (2007) noted that written records mainly concerned gentry and yeomen, but remarked that such opportunities were unlikely to be neglected by country folk generally. Eggs of many species were taken to eat, particularly seabirds (see Chapter 8) and ground-nesting birds (Chapters 5 and 7). Some seabird eggs were also used for industrial purposes in Britain. Even passerines' eggs were exploited. Macgillivray (1837–52) noted that the eggs of small birds were delicious, and he had often eaten those of Meadow Pipit *Anthus pratensis*, Wheatear *Oenanthe oenanthe*, thrushes and Corn Bunting *Emberiza calandra*, roasted in peat ashes as a boy. In the Victorian era collecting birds' eggs for the cabinet, particularly those of rare breeding species, was a fashionable pastime in Britain, which damaged the populations of some (see Chapter 10).

CHAPTER 2
Fowling methods

Before the use of shotguns became widespread, the main methods used to take birds involved the use of traps, snares, nets and bird lime (an adhesive substance that is spread onto a branch or twig). The sections that follow draw heavily upon Macpherson (1897) and describe briefly some of the principle methods and devices used, mainly in Britain and Europe. Variants of these methods and tools recurred worldwide.

It is difficult to discern any marked regional or specific patterns in the use or choice of the various devices. Snares and nets were perhaps the most widespread, and bird lime was most often used for passerines. But all these methods were widely employed, and use was probably dictated by the circumstances of the hunt; particular methods were also sometimes barred by landowners.

Fowling in ancient times

Clark (1948) considered that the main means by which prehistoric fowlers took birds were snares and simple traps, of types which still occurred over large areas of Eurasia and North America, or simply by clubbing them or catching them by hand when they could not fly. The use of nets is just as ancient, and taking wildfowl in clap-nets is widely illustrated in ancient Egyptian art (Houlihan & Goodman 1986); clap-nets and spring traps illustrated in Pharaonic times remain in use in Egypt today (Darby *et al.* 1977). The use of bird lime is also ancient, for it was certainly known in classical Greece (Thompson 1936). Many devices and methods were worldwide in distribution and recurred in widely differing cultures. Their occurrence in widely separated regions suggests that they were also devised independently.

Missiles were generally less important to early fowlers, but the Egyptians used throwing sticks to take wildfowl (see Plate 1) and Clark (1948) noted that club-headed arrows of the type used by modern circumpolar peoples for shooting birds have been found in prehistoric sites in Denmark. They are also illustrated in miniatures of men fowling in mediaeval manuscripts. Early modern fowlers in Europe generally used crossbows. Sometimes these were stonebows (known in Italy as ballistas), crossbows with several strings threaded through a pouch which launched a handful of lead or clay pellets. Single-shot slings were also occasionally used, particularly for small birds such as thrushes travelling in flocks (Clark 1948). Clark quoted a German source published in 1855 that around 1,200,000 Fieldfares *Turdus pilaris* were taken with slings annually in East Prussia. Whilst this seems slightly improbable, it may be that ballistas were meant.

Decoys

An important part of the paraphernalia of the fowler was the use of decoys and 'call-birds'. Call-birds were hung in cages around the trapping site, or placed in cages on high poles so that they could see congeners approaching and call them in. Other decoys were tethered or penned after being pinioned in the catching area, where their movement attracted others, or attached to a perch which could be moved by the fowler, making the decoys flutter into the air. This device was used for plovers (Haverschmidt 1943) as well as passerines. The use of call-birds was supplemented by a variety of artificial calls or by the fowler imitating bird calls.

Decoys were involved in the capture of virtually all the species regularly taken. Wildfowl and waders were frequently decoyed with stuffed birds (known as 'stales' in England) or artificial decoys of wood or metal painted to resemble the target species. Kear (1990) noted that the use of such models for attracting and killing ducks was mainly the invention of the North American Indians, for whom it was an ancient technique, and European settlers copied their use from that source. Indeed, although modern wildfowlers use them, the older literature for Europe rarely mentions decoys. But artificial wader decoys were certainly used in the past in Lincolnshire in netting Knots (MacPherson 1897).

Movement was also used to attract birds to the fowler. Markham (1621) recommended 'some twenty or thirty paces beyond your Netts, and as much on this side, place your Gygges, or playing wantons; being fastened to the toppes of long poles and turned into the winde, so as they may play and make a noyse therein … Birds will come and in great flockes to wonder and play about the same'. The gygges were like shuttlecocks, made of goose and swan feathers. A famous device used to attract larks was the lark glass or twirler. Records and descriptions of its use go back to at least the 16th century (see Chapter 9). Little Owls *Athene noctua* were also used as decoys in Europe, either in conjunction with call-birds, which created a fuss, or perched prominently to attract scolding groups of passerines into the catching area. Another form of such fowling in Europe was to use an Eagle Owl *Bubo bubo* to attract birds of prey or crows, which readily mob this species. They were used to take falcons for falconry and, in the mediaeval period, to attract kites, flying falcons at which was a highly regarded sport, to the falconer. In the modern period birds of prey and crows were shot over Eagle Owls, a form of persecution continued in France into the second half of the 20th century (Terasse 1964).

Another method of taking ground-roosting birds used a combination of light, bell and net, particularly in pursuit of Skylarks *Alauda arvensis*. In England this was known as low-belling (see p. 164) and, in the 16th century, the right to take larks thus was regarded as sufficiently valuable to be confined to the owners of land. At night lights and noise were also used to confuse wildfowl and waders and enable a close approach to catch them, a method used over a wide area of Europe, Asia and North America. Birds were either taken in nets or, in eastern North America and France particularly, enticed closer for shooting (see Chapter 5).

Snares and nets

Prehistoric fowlers probably made snares from vegetable fibres; the idea of making such traps seems likely to have originated with finding birds caught in such material. Arctic peoples used snares made from fine whalebone (Macpherson 1897). But in historic times the most common material used before fine woven wire became readily available was woven or plaited horse-hair – fine, strong and widely available. Mitchell (1885) noted that

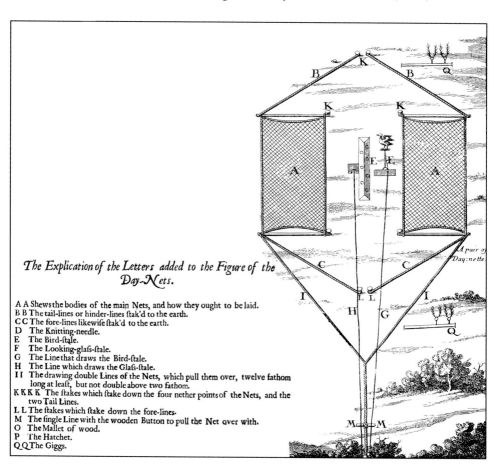

The Explication of the Letters added to the Figure of the Day-Nets.

A A Shews the bodies of the main Nets, and how they ought to be laid.
B B The tail-lines or hinder-lines ſtak'd to the earth.
C C The fore-lines likewiſe ſtak'd to the earth.
D The Knitting-needle.
E The Bird-ſtale.
F The Looking-glaſs-ſtale.
G The Line that draws the Bird-ſtale.
H The Line which draws the Glaſs-ſtale.
I I The drawing double Lines of the Nets, which pull them over, twelve fathom long at leaſt, but not double above two fathom.
K K K K The ſtakes which ſtake down the four nether points of the Nets, and the two Tail Lines.
L L The ſtakes which ſtake down the fore-lines.
M The ſingle Line with the wooden Button to pull the Net over with.
O The Mallet of wood.
P The Hatchet.
Q Q The Giggs.

Figure 2.1. *Plan of the layout of a clap-net, from Ray 1678. Such nets were used by bird-catchers, with scarcely any change in design, from ancient Egyptian times right up to the 20th century.*

a single strand of horse-hair was strong enough to hold a lark, and two were sufficient for Snipe *Gallinago gallinago* and Teal *Anas crecca*, but heavier snares were used in springes (horsehair-triggered traps; see p. 130). Many species were taken in snares, from wildfowl and waders to gamebirds such as grouse and many passerines.

Nets came in great variety, from small hand nets used to take larks at night to flight-nets set along the shore for waders and wildfowl, which could be up to 800m long. But the three most important basic types were clap-nets, drag-nets and flight-nets.

Clap-nets came in several types but the principle on which they worked was the same in all. Ray (1678) illustrated it very clearly (Figure 2.1). Two panel-nets were laid out on the ground with a space between. The panels were mounted on poles hinged at the base and connected to the trapper by a long pull-cord. Call-birds and decoys were caged or tethered by the nets, just clear of their throw, and bait often scattered on the ground between. As birds alighted the cord was pulled, throwing the nets over. This device had a very long history, with a design virtually unchanged from ancient Egyptian times. It also had a world-wide distribution, although it was only introduced into South Africa and Australia in the 19th century (Macpherson 1897). Macpherson noted that the people most skilled in its use were the Italians, and elaborate permanent constructions were built for netting birds on Italian estates and in Germany. Geese were taken with clap-nets throughout Russia and Siberia and also in the Netherlands. Waders were taken in single panel-nets which worked on the same principle (see Chapter 7).

The dimensions of these nets did not vary greatly, except that the Russian nets used for geese were up to 36m long and 2m deep. Mitchell (1885) gives panels 11m long by 2.5m deep for the clap-nets used by the Lancashire bird catchers, Pennant (1776) gave 12m by 2.5m for the nets used for taking Ruff *Philomachus pugnax* and Payne-Gallwey (1882) and Haverschmidt (1943) gave 14m by 2.5m for nets for catching plovers.

The drag-net, also known as a trammel net, another ancient device illustrated in ancient Egyptian art was, as its name suggests, pulled over the ground and dropped on birds. It was used at night in pursuit of birds such as larks and partridges. Drag netting was increasingly restricted from the mid-19th century by the rise in game preservation, particularly of partridges for driving, which was incompatible with drag-netting at night. Partridges were also taken in tunnel-nets, which were similar in design and function to duck decoy pipes (see p. 103), with netting wings to guide the birds; these were attracted by bait, and gently pushed towards the tunnel by a fowler dressed as a cow (*i.e.* with a cow's head and skin draped over the shoulders; see Figure 6.4, p. 103).

Flight-nets were similar in function and design to the modern mist-net, although those set for waders and wildfowl may have been rather heavier (Figure 2.2). They were principally used for wildfowl and waders in estuaries in Britain, and for passerines. Very long runs were often set for wildfowl. Small passerines were often taken with what were known as spider nets (see Chapter 9). In Germany and the Netherlands very large flight-nets were used to trap Starlings *Sturnus vulgaris* roosting in reed beds. These nets were up to 30m long and 21m deep, stretched between long poles and worked with lines and pulleys. They were suspended outside the reed bed with high sidewalls of netting to keep the birds from escaping; the fowlers drove the roosting birds into the nets after dark (Macpherson 1897).

Figure 2.2. *A flight-net, in use on the Lancashire salt marshes. From Mitchell (1885).*

Bird lime

Bird lime was manufactured from a number of natural bases that had the necessary viscosity. Mistletoe berries were commonly used in Europe, but the bark of the holly was the traditional base constituent in Britain. Bird lime based on linseed oil could also be purchased from chemists in England.

Bird lime was mainly used to take passerines, although partridges and Lapwings were taken with it, and Japanese fowlers took ducks with limed threads. It might be spread direct on to prepared twigs on bushes or trees but was more usually used on specially prepared lime sticks, which the fowler carried in sheaths made of hollowed wooden tubes. Such sticks were used in many situations, but the principle behind their use varied little. They were placed in tempting positions for birds to perch, and a combination of calls and decoy birds attracted the target species to alight. They had little chance of escape once they touched the lime sticks, which were often so delicately balanced that they dropped to the ground when birds alighted. Little Owls were also used in Europe to attract mobbing parties of song birds.

Gregarious birds were also taken with limed threads; these were attached to a captured bird, which was then released to join a feeding flock, trailing the thread. This inevitably entangled others that came into contact with it. Larks were also taken by being driven gently up to limed strings stretched across a suitable field.

Guns and stalking horses

The use of the gun for sporting purposes starts to appear in the records in the first half of the 16th century. But guns had been in at least occasional use for some time by then, as regulations on their use date back to the 15th century in Scotland (see Chapter 3). Early guns were cumbersome, inefficient and expensive. In Britain until long into the 17th century, shooting with the gun was regarded as a sport for the lower orders only, and Markham (1621) advised his readers to shoot their birds on the ground and aim to obtain as many as possible at one shot. Following Markham's advice stalking horses, trained to be steady despite gun shots fired under or over their necks, were used to approach closely to gamebirds and wildfowl (Figure 2.3). Where birds became familiar with stalking horses, stalking deer, made from canvas with real antlers, or artificial oxen were used. Stuffed horses were also sometimes used and Brusewitz (1969) illustrates a splendid device resembling a pantomime cow, with hollow horns through which the gun was poked to fire. Chapman & Buck (1893) gave a detailed account of the use of the stalking horse in the marismas of Andalucia. Professional fowlers used them in pairs. The horses had a short rein running from the head to one foreleg so that they gave the appearance of grazing (Figure 2.3). Some considerable bags were made. Chapman & Buck also once took a gunning punt and big gun to this area, but it was impounded by Spanish Customs as a 'warship'. It took a season to extricate it from the morass of red tape, after which they found that it was virtually useless as the fowl would not allow a close enough approach to offer a shot.

Steady improvements in sporting guns were made over the 17th and 18th centuries, particularly with the development of the flintlock in the early 17th century, and the sport of shooting flying birds was introduced. An early enthusiast was Frederick William I of Prussia. He shot partridges two or three times a week in the early 18th century '...with a small retinue of servants to load his guns and accompanied by falconers and gamekeepers he sometimes returned home with fair-sized bags – 160 partridges on one occasion. Misses were, of course, legion and an eye-witness assures us that the royal gun often fired 600 shots or more in a day in partridge country' (Brusewitz 1969). Hunting on this scale in Europe was the preserve of princes and the aristocracy, and it played its part in projecting the status and power of these elites (Blanning 2007). Sportsmen walked towards their game with pointers and setters, used to find and set the game, *i.e.* induce it to stay put until the gun was ready. The sportsmen then flushed and shot at the birds over the dogs.

Muzzle-loading flintlock fowling pieces reached their fullest development in the second half of the 18th and early 19th centuries, and quality guns of this period are beautiful weapons to handle. But they always had considerable drawbacks. They were not always certain of firing and were slow to load and in firing, with a distinct pause between pulling the trigger and the charge exploding. Powder was of very variable quality and there were no standard shot sizes. As is evident from innumerable sporting prints of the 18th and early 19th centuries (Plate 4), most shooting was at birds flushed and flying away where the gun could be simply held on the target. But this was always a slightly risky business. Hastings (1981) had personal experience of shooting with old flintlocks and he is worth quoting.

The prudent man with a muzzle-loader (although many used dry grass for wadding and measured out their powder in the bowl of a clay pipe) had spring measures fitted into flask and pouch so that he could pour the correct load of powder and shot. But it was all wildly

(a)

(b)

Figure 2.3. *Stalking horses; variations on a theme. (a) – in use stalking wildfowl in the 17th century, from Blome's* The Gentleman's Recreation *(1686); (b) – stalking wildfowl in the marismas of southwest Spain, from Chapman & Buck (1893); these are known as cabresto ponies.*

experimental; a matter of personal opinion on what suited the gun best. As powder was unpredictable in its performance, and shot was uncertain in size, every sportsman fancied himself his own best judge.

And then:

The hazards of the sport with a muzzle-loader were formidable. In the excitement of the chase sportsmen fired their ramrods after departing game. In the hurry to reload after a hangfire, which was common, there was always the danger of looking down the barrel to see what had gone wrong, and catching a charge in the face. If powder was poured into the barrel over a winking spark, flasks were liable to blow up ... Double-barrelled guns, an early innovation of the gunmakers, were doubly dangerous. It was so easy to double charge one barrel. Shooters were recommended to protect themselves by leaving the ramrod in the barrel they had charged before pouring powder into the other.

Overcharged barrels were liable to burst, which is why shooters look so awkward in prints, with the leading hand held back close to the breech where a burst was less likely. With double-barrelled muzzle-loaders it was not unknown for the unfired barrel to go off when the other was being reloaded, a hazard for which Manton designed a safety lock that would not fire when the gun was vertical (Chute 2002).

Over the first 70 years of the 19th century there were steady improvements in sporting guns, through a progression from muzzle-loading flintlocks to muzzle-loaders fired by percussion caps to breech-loaders firing pin-fire cartridges, in which the firing pin stuck out of the cartridge and fired when hit by hammers on the gun. These led via breech-loaders firing centre-fire percussion cartridges to the hammerless ejector gun used today (Boothroyd 1985). All sorts of patents and designs emerged and were tried along the way, but these were the main steps. They were paralleled by improvements in ammunition, particularly the standardisation of loads and shot sizes, and the development of smokeless nitro powders to replace the old black powder. The aim of such improvements was constant, to improve the speed, certainty and safety of firing, and the development of breech-loaders and percussion cartridges, particularly, made large bags of driven game possible. With them emerged the whole battery of strict preservation, intensive 'vermin' destruction, and large-scale driving of reared birds.

CHAPTER 3
Regulation and seasons

Game laws and poaching

In England, regulations or restrictions governing who could take gamebirds date back to at least the 14th century. The method used from the end of that century was a property qualification; those who did not possess property of a certain value were forbidden to hunt game, even on their own property (Munsche 1981). Conversely, under the Game Act of 1671, qualified persons could hunt game anywhere, a potent cause of resentment in the countryside.

There were also regulations on hunting methods. As early as 1427 an Act of James I of Scotland laid down that there was to be no shooting with hail-shot or hand gun within 600 yards of a heronry (Gurney 1921), and three other Acts were passed in Scotland between 1493 and 1599 for the protection of herons (Baxter & Rintoul 1953). In England a regulation of Henry VII in 1504 forbade the taking of Grey Herons *Ardea cinerea* except with hawks or a longbow (Stubbs 1910b, Gurney 1921). This was not terribly effective, as entries in the Le Strange household accounts, for example, record herons, together with Mallard *Anas platythychos*, Bittern *Botaurus stellaris*, Common Cranes *Grus grus* and Great Bustards *Otis tarda*, being shot with the crossbow in the 1520s (Gurney 1834, Gurney 1921).

Large areas of the countryside in the mediaeval and early modern periods were waste and common, which comprised at least 25% of the agricultural area of England and Wales. The nearest modern equivalent to such habitats are the categories 'rough grazing' and 'common rough grazing' in the June census of agricultural statistics, which still form around 14% of agricultural land. Of greater significance than its larger area in the past, however, was its distribution. Whilst today rough grazing habitats are highly concentrated in the uplands, in the period before the mid-18th century all lowland parishes had significant areas of waste and common, which were largely managed under rights in common and provided important resources in grazing, fuel and other materials to their rural communities (Shrubb 2003). There, too, wild plants could be gathered and birds caught for food by those prepared to take the trouble (Thirsk 2007).

In some cases at least, certain species were always reserved for the Lord's use. Thus in the Earl of Northumberland's Household Book of 1512 Great Bustards were only to be taken 'for my Lordes owne Mees [mess or table] at Pryncipall Feestes Ande noon outher tyme Except my Lordes commaundment be otherwyse'. Cranes were similarly reserved for the Lord's table at Christmas and other principal feasts as were plovers, herons, Mallard, Quail *Coturnix coturnix* and Woodcock *Scolopax rusticola*. Sixteen other species were subject to similar regulations, and these instructions also laid down the price to be paid, sometimes adding the qualification 'so they be good' (*i.e.* so long as they are of sufficient quality) (Nelson 1907, Woolgar 1999).

But generally speaking, no laws prohibited the trapping of birds as long as they were not taken on the Lord's demesne, in his warren or his park (Thirsk 2007). Theoretically this was changed by the Game Law enacted by Richard II in 1390, which restricted hunting to those with property valued at a minimum of 40 shillings per year (Almond 2003). How effective this was is questionable. Almond noted that poaching by all classes, including the clergy, was rife, whilst Markham (1621) relied on peasants and labourers to teach him about the trapping of birds. It is clear also, from the wide range of species offered for sale that fell outside the definition of game, that the activities of rural fowlers supplied urban markets (see Table 1.4). Indeed Munsche (1981) noted that in 17th-century London, Winchester, Oxford and other large cities, game was bought 'in quantities large enough to indicate a supply which could not have come from any but professional fowlers or country folk'.

Much of this changed in the late 17th and early 18th centuries, starting with the Game Act of 1671. This was the main basis of the increasing thicket of game laws passed in the 18th century, up to 1831. The purpose of all this legislation was not to protect game (defined for the purpose of the Game Laws as partridges, pheasants, grouse and hares) but to render hunting this game the exclusive privilege of the landed gentry. The Act aimed to achieve this by three devices. It forbade all persons with freeholds of less than £100 per year or leaseholds of less than £150 per year to hunt game, even on their own land. Conversely, qualified persons could hunt anywhere, as the concept of game as the private property of the owner on whose land it was found did not exist. Secondly, it permitted the seizure of guns, nets, dogs, snares and other hunting equipment from unqualified persons. Thirdly, it banned the public sale of game by poulterers, a measure first brought in 1603 but long disused (Munsche 1981).

The efficacy of such legislation in protecting the landed gentry's privilege seems doubtful when, between 1671 and 1831, 53 assorted Acts and regulations, some admittedly short-lived, were brought into force to bolster it. One major obstacle to its implementation was that banning the sale of game proved unenforceable, and the demand for game from urban populations led to an increasing scale of profitable commercial poaching to serve it (Munsche 1981). To an important extent this was enabled by the great improvements in roads, and the regularity of coach services in the second half of the 18th century, particularly with the introduction of the mail coach service from 1784. Munsche noted that

> *in addition to passengers and mail, coaches carried perishable agricultural produce, including game. Stewards sent game to their masters; gentlemen sent it to their friends; and higglers* [itinerant poultry dealers] *sent it to their customers. For the drivers and*

guards of these coaches it was only a short step from carrying game for others to dealing in game on their own behalf. Assisted by the keepers of the inns where they stopped, coachmen soon became as important in the operation of the game trade as the higglers themselves.

They operated outside the established structures of the poultry trade and, in these activities, were subject to no effective regulatory authority. Coach drivers and guards became an essential conduit between increasing numbers of commercial poachers and the London and other markets.

Mayhew (1861–62) made the same point. He found that about 16% of the game (as defined by the Game Laws) on sale in London was sold by street hawkers, who obtained it directly from poachers. Street hawking of poached game did not long survive the passing of the 1831 Game Reform Act. Until that Act legitimised the trade, the salesmen in Leadenhall market used the same source, Mayhew noting that 'the purveyors for the London game market – I learned from leading salesmen in Leadenhall – were not then, as now, noble lords and honourable gentlemen, but peasant or farmer poachers, who carried on the business systematically. The guards and coachmen of the stage coaches were the media of communication, and had charge of the supply to the London market'.

By the early 19th century the scale of commercial poaching and the measures taken by landowners to combat it amounted to a minor civil war in the countryside, and led Parliament, in the Act of 1831, to end the property qualification and declare game to be the property of the landowner on whose land it was found. Licences to kill and trade in game were also introduced and, in 1870, gun licences.

The growth of game preserving

An important innovation in game shooting in Britain was the introduction from the Continent of the battue (Figure 3.1), where game from a large area was driven into an extensive enclosure of cover, usually surrounded by netting or a wall to prevent it running out, and then shot by guns and beaters walking through in line. Shooters used several guns and loaders to facilitate rapid firing. The battue came into Britain in the late 18th century but was regarded askance by many traditional sportsmen, not least because the management of estates for this purpose restricted their activities (see p. 44). Nevertheless, improvements in guns and marksmanship generated a competitive interest in the bags sought and obtained. To satisfy this, game was increasingly bred and protected on estates, which ensured the battue's continuance. The modern practice of driving game evolved from it.

Parliamentary enclosure (enclosure of common lands and open fields by act of parliament) from the mid-18th century greatly assisted this trend for it converted extensive areas of land managed under rights in common, where game could hardly be preserved, to land held by individual proprietors, where it could. Enclosure also led to a significant decline in the number of small freeholders and an expansion of large estates (Shrubb 2003). With the close preservation of game, old practices, such as the netting of larks and other songbirds, Woodcock and plovers came to be increasingly restricted to a declining area of open or common land, as the 19th century avifaunas make clear; Macpherson (1897) made much the same point for Italy and Germany. Such activities were incompatible with the rearing of large stocks of game for driving.

Figure 3.1. *A battue in Windsor Great Park. This was the occasion of King Haakon of Norway's visit to Britain in 1905. King Edward VII is on the left, with Haakon just to his right, and the future George V is shooting downward in the foreground. From a contemporary press report.*

That also curtailed the right of qualified sportsmen to pursue game anywhere. Game increasingly came to be regarded as private property, bred and preserved in protected coverts by the landowner, where it could be hunted only with his permission (Munsche 1981). This was a change which led to marked increases in gamebird populations, but it was not a development welcomed by many traditional qualified sportsmen. Peter Hawker in his diaries not infrequently boasted of shooting in other landowners' preserves and occasionally running away from their keepers to prevent them from formally warning him off from doing so, which would have left him liable to legal action if he persisted. In his diary for September 1812 he describes in some detail just such an encounter with one Squire Jones at Bradford in Wiltshire, which generated a lengthy and fairly acrimonious correspondence.

Hunting seasons

As noted above, the English Game Laws were not concerned with the conservation of game species. That required separate legislation, and the first statute to establish seasons for taking gamebirds was passed in 1762. In 1773 the present seasons for taking gamebirds were established, with grouse starting on August 12th, partridges on September 1st, and pheasants on October 1st.

Scottish legislation differed markedly, with close seasons introduced in the early 15th century. Thus an Act of James I in 1427 set close seasons for Red *Lagopus lagopus scotica* and Black Grouse *Tetrao tetrix*, Grey Partridges *Perdix perdix* and plovers from the beginning of Lent to August. Subsequent Acts protected the nests and eggs of partridges and forbade the killing of any edible fowl in moulting time in 1457, extended the close season for Red Grouse to Michaelmas in 1555, and confirmed the close season for Black Grouse, forbade the sale of Ptarmigan *Lagopus muta* and shortened the close season for Red Grouse to July 3rd in 1599. That close season was reduced again to June 21st in 1707. The Capercaillie figured only once in such legislation, when an Act of 1621 forbade their purchase or sale, under a penalty of £100 (Scots pounds). Ptarmigan and Quail were similarly protected under this Act, and a close season was introduced for both species in 1685 (Baxter & Rintoul 1953, Gilbert 1979). In 1772 legislation and seasons were brought into line with those in England, and this was confirmed by an Act of 1832, which applied the provisions of the Game Reform Act 1831 to Scotland.

Wildfowl, principally Mallard, Teal *Anas crecca* and Wigeon *Anas penelope*, were excluded from the game legislation, but open and close seasons had already been introduced for these birds (from as early as 1534). As far back as the reign of King John, concern had been expressed in England about the impact on native populations of taking large numbers of flappers (young ducks unable to fly properly, or adults in moult) in areas such as the Fens. An Act of 1534 aimed to remedy this. Its preamble stated that ducks, mallards, wigeons, teals, wild geese and 'divers other kinds of wildfowl' were suffering because

> *divers persons next inhabiting in the countries and places within this realm, where the substance of the same wildfowl hath been accustomed to breed, have in the summer season, at such time as the said old fowl be moulted, and not yet replenished with feathers to fly, nor the young fowl fully feathered perfectly to fly, by certain nets and other engines and policies, yearly taken great number of the same fowl in such wise that the brood of wildfowl is almost thereby wasted and consumed.*

The Act had two parts, the first prohibiting the taking of fowl between May 31st and August 31st, the second prohibiting the taking of eggs of certain birds, on pain of a year's imprisonment and a scale of fines ranging from 20d for any crane or bustard egg down to one penny for ducks (Gurney 1921, Darby 1934). The first part of this Act was repealed in 1550 on the grounds that benefit was 'thereby taken away from the poor people that were wont to live by their skill in taking of the said fowl, whereby they were wont at that time to sustain themselves, with their poor households' (Darby 1934).

Restrictions on the times when wildfowl could be taken were reenacted in 1710, when the close season was fixed as July 1st to September 1st, which was clearly aimed at restricting the taking of flappers. The close season was extended in 1739 to the period between June 1st and October 1st (Darby 1934).

Efficacy of game laws and seasons

It is unlikely that such legislation was ever very much respected. The evidence of the household accounts of English gentry families in the 16th and 17th centuries suggests that there was little concept of a close season. Figure 3.2 summarises the months in which birds of three groups, wildfowl, game and waders, were paid for in three gentry households. It shows a pattern strongly influenced by the seasons, with the largest numbers recorded in the autumn and winter and a pronounced trough in February and March. Such a pattern at least partly reflected the status of many of the species concerned as migrants and/or winter visitors, arriving in autumn, with large numbers through the winter and a rapid departure in the early spring. But this spring trough also partly reflected the observance of Lent, when no birds were consumed in at least some households (Stone 2006). There was a sustained level of activity throughout the period between April and August, suggesting

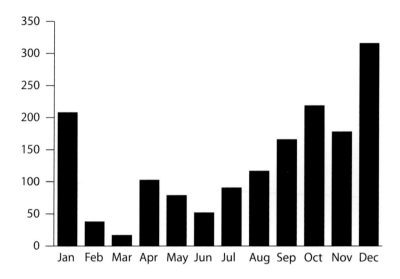

Figure 3.2. *Numbers of birds of three groups – wildfowl, gamebirds and waders – paid for in three English households in the 16th and 17th centuries by month. Data from Gurney (1921) and Bourne (1999).*

that little heed was paid to the close season. Gilbert White was recording the taking of flappers on the ponds in Woolmer Forest in July 1773 (White 1789, in letter 39 to Pennant). Hawker (1893) made a visit to Norfolk in mid-July 1816 to fish and shoot young wildfowl. His diary records that

> *the young wild fowl shooting was most capital. We killed large numbers of almost every kind of sea and marsh birds, interspersed with occasional good shooting at leverets and rabbits, young snipes, plovers etc. The only birds, however, I had not killed before were the crested grebes and the shoveller ducks, with which I had, one day in particular, most excellent sport.*

Such records suggest that the taking of flappers remained common practice.

European game legislation

A rather similar situation existed in parts of Europe. For example, Table 3.1 summarises the regulations imposed to control the taking of birds in various parts of Germany from the 16th to the 18th centuries. The constant repetition of these orders argues clearly that they were never very effective. Hobusch (1980) also made the point that many of these regulations protected birds only in the interest of hunting, and close seasons were curtailed over time in the interests of increased hunting. Stocks of game were considerably reduced and, by the mid-19th century, were at their lowest. By the end of the century, representatives of animal protection and conservation organizations were raising serious concerns.

The 19th century proved a difficult period for many quarry species. The collapse of aristocratic privilege in Europe after the political upheavals of the late 18th and mid-19th centuries certainly contributed to this, since aristocratic elites lost the exclusive right to hunt; the resulting free-for-all led to steep declines in many game animals (Hobusch 1980).

But in any case, close seasons had more often been curtailed than extended, at a time when guns were rapidly improving in efficiency. This resulted in a pattern of excessive shooting and trapping in the early breeding season or in the spring migration period, when bird populations tend to be at their low point. Thus, in Britain, with the passing of the Game Reform Act in 1831, wildfowl lost the protection given to them by earlier statutes (Marchant & Watkins 1897); this lack of any properly enforced close season had, by the second half of the 19th century, come to have a significant impact on British wildfowl populations. Species such as Teal, Pintail *Anas acuta* and Shoveler *Anas clypeata* were widely recorded as declining in the 19th century avifaunas, with the latter two recorded as uncommon or scarce. Nelson (1907), for example, wrote of the Pintail that 'half a century ago, the Pintail was a numerous species on the Tees, where, as Geo. Mussel tells me, it was greatly sought after by the professional gunners, who would not trouble with other fowl if they could get the Pintail and, as it was most plentiful in May, and no restrictions were at that time placed on shooting, great numbers of this delicious duck were procured and brought into market'. Garganey *Anas querquedula*, too, were frequently recorded as being shot or taken in decoys in April and May. Such problems were European in scale. Brusewitz (1969), for example, describes the scale of hunting of wildfowl at all seasons in northern Europe and Scandinavia,

Date	Place	Provisions
July 2nd 1572	Mecklenburg	Close season fixed from March to July 25th for wildfowl, bustard and partridge
February 4th 1575	Weimar	Close season fixed from March to August 24th
March 6th 1582	Brandenburg	Shooting of swans forbidden
1606	Königsberg	Close season fixed from March to August 24th
February 1st 1622	Brandenburg	No eggs to be taken from nests or breeding birds disturbed
May 12th 1670	Brandenburg	Taking of eggs and destroying nests forbidden; repeated June 9th 1677, March 18th 1680, April 5th 1698, April 10th 1704.
August 25th 1686	Brandenburg	Nightingales protected
June 17th 1698	Königsberg	Nightingales protected
August 20th 1743	Magdeburg	Limit set on number of larks to be taken in nets
November 9th 1705	Prussia	Close season for snipe and duck set at May 1st to July 1st; repeated March 11th 1713 and April 8th 1715
April 10th 1709	Prussia	Close seasons for all game, including nesting season of birds, to be observed
December 22nd 1723	Prussia	Wild duck and swans to be protected when nesting, but geese, cranes, herons and wood pigeons can be shot at any time
October 3rd 1743	Prussia	Close seasons curtailed further
April 18th 1755	Mecklenburg	All close seasons and rules for protection of game cancelled

Table 3.1. *Regulations governing the hunting of birds in Germany from the 16th to the 18th centuries. Data from Hobusch 1980.*

which led to considerable population declines in the 19th century (see Chapter 5). The London markets imported ducks from even further afield; northern grouse populations were also decimated (see Chapter 6).

Waders were similarly affected, particularly species such as Ruff *Philomachus pugnax* and Black-tailed Godwit *Limosa limosa*, whose status had already been seriously undermined by land drainage. Stevenson (1870) noted that the Black-tailed Godwit, formerly an abundant breeder in the Netherlands, had declined markedly there through over-exploitation. Brusewitz (1969) observed that the wader most affected by civilization in northern Europe was the Great Snipe *Gallinago media*. Not only had the increase of cultivated land rapidly restricted its breeding habitat but it was also shot and trapped in large numbers at its display grounds in spring, and in huge numbers on autumn passage, when it was noted as a particularly easy target (Chapter 7). Egging was also a problem for waders, especially Lapwings *Vanellus vanellus*, for which heavy exploitation of their eggs and spring netting and shooting had significantly depressed populations in Britain and Europe by the second half of the century (Shrubb 2007).

What legislation there was governing the taking of birds up until the late 19th century was largely concerned with game and wildfowl. There were few controls of any sort on the exploitation of other birds, despite the scale on which species such as larks and thrushes were taken and the very wide range of species which were persecuted, for food, for the cagebird trade, for their plumage or for specimens for display cases. Many of the seasons discussed in the legislation for taking songbirds for food, for example, were not so much legal devices as periods in the birds' annual cycles, when abundance made trapping profitable. Nevertheless, some legal restrictions were promulgated. Macpherson (1897) noted that the trapping of thrushes had by then been made illegal in Denmark and, in Germany, it was forbidden to snare resident thrush species. This was ineffective, since residents were still caught in snares set for migrants. The use of nets and mirrors for taking larks was also regulated by law in Germany and, when Macpherson wrote, the use of the *lerchenstreichen* (see p. 164) had been curtailed or forbidden. In Italy, the legal season for trapping birds was expressly limited to the period encompassing the autumn migration, but it was little regarded. Macpherson noted that Bergamo, Verona and Brescia were the chief centres of fowling, and the issuing of fowling licences was an important source of tax revenue; Bergamo issued 1,133 licences to net and 4,824 to shoot annually, although many peasants operated without legal consent.

In Britain, although there were no equivalent regulations, considerable protection was provided to many common birds of the countryside by the effect of enclosure and the increasingly close protection of game, which severely limited access to large areas of the countryside. Neither landowners nor gamekeepers were likely to tolerate the activities of bird-catchers in their fields and coverts. As already noted, this was often commented upon in the 19th-century avifaunas. Poaching was rife, but poachers were after gamebirds, not songbirds.

The plumage trade (see Chapter 11) was responsible for the most severe slaughter of birds at this time. This unrestrained exploitation saw birds destroyed in their millions, largely in the breeding season when plumage was at its best. Herons and egrets, with their beautiful plumes, colourful species such as pheasants, kingfishers and birds-of-paradise, and seabirds with their white plumage, which became highly fashionable in the second half of the 19th century, were particularly at risk.

Seabirds

Seabird colonies were also subject to mindless shooting from boats, laid on as sport for the entertainment of parties of tourists; Kittiwakes *Rissa tridactyla* were particularly badly affected, and their wings were taken to decorate ladies' hats (Chapter 11). This practice was being indulged in as early as the first decade of the 19th century. Hawker's (1893) diary for July 9th 1811, for example, records that he 'went in a boat to the Needles for rock-shooting and killed among other birds a Cormorant *Phalacrocorax carbo*. My killing the latter bird was considered great sport, as the boatman and other people informed me that it was the first they had seen dead the whole season; for, although every shooting party had tried every way for them, the Cormorants were so difficult of access, and (even when within reach) carried away so much shot, that none had been killed'. On the 10th Hawker landed on the Needles; he shot two more Cormorants, and then 'went to have a few hours' pastime under the rocks, but found the birds so very wild that I despaired of getting shots, but by dint of perseverance killed 5 Puffins, 2 Razorbills and 3 Willocks [Guillemot, *Uria aalge*]'.

Seabirds, of course, were widely exploited for food, but the North Atlantic communities that depended on seabird fowling to any extent tended to have evolved traditional methods for protecting stocks from over-exploitation (see Chapter 8). But this was not invariable, as Diamond's (2005) account of some of the Pacific islands, particularly the extreme case of Easter Island, indicates. Seabird colonies were an important source of revictualling for ships undertaking long journeys, across the Atlantic for example; captains often planned such long-distance voyages so that advantage could be taken of seabird colonies for food, an important cause behind the extermination of the Great Auk *Pinguinus impennis* (Greenway 1967).

The obvious over-exploitation of birds in the second half of the 19th century led to increasing demands for laws to conserve and protect them. It is worth making the point that it was reaction to the excesses and barbarities of the plumage trade that laid some of the important foundations of the modern conservation movement, for both the Audubon Societies in America and the RSPB in Britain arose directly from opposition to that trade. But before that, the concern generated by declines in seabirds and wildfowl led to a series of Bird Protection Acts being passed in Britain between 1869 and 1896, which have been updated and reinforced twice since.

The original Act of 1869 was confined to seabirds. Its preamble stated that 'the seabirds of the United Kingdom had greatly decreased in number, and that it is expedient therefore to provide for their protection in the breeding season'. The Act made it an offence to kill or wound, or attempt to do so, any of 33 species of seabird, or to use any boat, gun or net or other device for that purpose between April 1st and August 1st in any year. The Act specifically exempted young seabirds unable to fly, presumably so that communities dependent on seabirds for food could continue to use this resource (Marchant & Watkins 1897).

This Act was succeeded by further Acts passed in 1872 and 1876. That of 1876 was specifically entitled 'An Act for the preservation of Wild Fowl', and its preamble stated 'the wildfowl of the United Kingdom, forming a staple of food and commerce, have of late years been greatly decreased in numbers by reason of their being inconsiderately slaughtered during the time they have eggs and young'. The Act fixed a close season from February 15th to July 10th, and not only forbade the killing or wounding of species covered by the Act in this period, but made it an offence for any person to have recently taken birds of these species in their possession. The evidence from the county avifaunas of the time is quite clear; many wildfowl and wader populations started to recover with the passing of the Act. This Act was replaced in turn by four more, passed in 1880, 1881, 1894 and 1896 and collectively known as the Wild Birds Protection Acts 1880–1896. These remained in force until superseded by a new Act in 1954.

CHAPTER 4
Herons, spoonbills and cranes

From at least the mediaeval period until the end of the 17th century, herons and cranes were valued in Britain and Europe as delicacies for the table and were in great demand for feasts and other occasions. The Grey Heron *Ardea cinerea* and Common Crane *Grus grus* were also valued as quarry for falconry. This taste for herons (and for storks and flamingos) was widespread throughout the Old World. Macpherson (1897), for example, describes methods of trapping them from Egypt, India and Japan as well as Britain and Europe. In parts of India, herons and egrets were also trapped for use as decoys when taking wildfowl in clap-nets, and Macpherson quoted A. O. Hume that hundreds of these birds could be seen tethered about every fishing village in Sind. People in this part of India also farmed egrets for their plumes (see Chapter 11). The use of herons as decoys to encourage wildfowl within range of clap-nets occurred in other cultures. All the heron species considered in this section were so-used by ancient Egyptian fowlers, and the practice is frequently illustrated in ancient Egyptian art (Houlihan & Goodman 1986). Markham (1621) advocated using Grey Herons for the same purpose in the 17th century.

Five heron species were exploited regularly for food in Britain and Europe: Bittern *Botaurus stellaris*, Night Heron *Nycticorax nycticorax*, an egret species, probably Little Egret *Egretta garzetta* (though this species is by far the likeliest candidate, this was never explicitly stated and it remains an assumption), Grey Heron and Spoonbill *Platalea leucorodia*. These birds and Common Cranes figured prominently in the provisions listed for major feasts and in household accounts (Table 1.3; see p. 20), and were commonly available from poulterers' stalls, although they were always expensive. Gladstone (1943) noted that Bitterns and Grey Herons were still offered in the markets into the early 19th century, but the other species disappear from such records by the 17th century. Macpherson (1897) also noted that many Purple Herons *Ardea purpurea*, Night Herons and Bitterns were imported into England in his day as pets or aviary birds, mainly as immatures and presumably taken before they could fly. In the early 19th century Little Bitterns *Ixobrychus minutus*, mostly adults snared on the nest, were sent alive from Holland to Leadenhall Market in May and June, although they seldom lived long (Lilford 1895)

Some culinary observations

One point of considerable interest is what fish-eating birds such as herons actually tasted like. The comments I have found in the 19th-century avifaunas give contrasting opinions. Bitterns appear not to have been fattened up in the pens known as stews, prior to heading to the pot (see below). A surprising number of the authors of these avifaunas had eaten them, and agreed with Pennant (1776) that they had much the flavour of a hare and nothing of the fishiness of the heron; Lilford (1895) described the flesh as excellent. Yarrell & Saunders (1885) noted that the flesh of the Spoonbill, although dark, was said to be of good flavour and without any fishy taste, and that the species, in common with its allies, would feed on any sort of offal when kept in captivity. Ticehurst (1932) also noted, from his own experience, that young Spoonbills were not bad eating. Of the Grey Heron, Hele (1870) considered the flesh to be very good eating, with a flavour somewhere between hare and goose, but Ticehurst (1932) found the one he tried very poor meat. Muirhead (1895) quoted James Smail as stating that that even when plump, dressed and roasted to perfection, the flavour of Grey Heron was so fishy as to be quite unpalatable; however, the young birds, when taken from the nest and properly stewed (*i.e.* fed in pens) were rather good. Bolam (1912) noted that young herons, when well-roasted, made by no means a bad dish, but Vesey-Fitzgerald (1969) found the one he tried loathsome.

Preparation seems to have been important, and instructions for cooking herons stressed the importance of not breaking the bones. In the 17th century herons were roasted larded in 'swynefat' and eaten with ginger; they were also made into puddings (Lowe 1954). Simon (1952) quoted Elyot's *Castle of Helth*, published in 1539, that 'all these fowls [*i.e.* herons, Bitterns and Spoonbills] muste be eaten with much ginger or pepper, and good olde wyne drunk after them'. It seems clear that the practice of penning herons in stews and fattening them on ox liver, curds and similar foods to sweeten the flesh made them palatable. This begs the question of why one would bother in the first place. But in the past the landed gentry depended almost entirely upon home supply to provision their considerable establishments, and such birds were regarded as part of the produce of the estate. Self-sufficiency was a major aim of household management (Southwell 1870–71, Gurney 1921, Thirsk 2007). In the mediaeval period, at least, display was also important, and the consumption of herons was an important mark of status.

The Bittern

The Bittern *Botaurus stellaris* was a widespread and reasonably numerous species in major wetlands from Britain right across Europe before the 19th century. The species is present in the Saxon archaeological record in Hampshire (Yalden & Albarella 2009), and was recorded in the household accounts of the Earl of Oxford in Essex in the early 14th century as being sent from his manor of Great Bentley (Woolgar 2006), which suggests a breeding site in the mediaeval period, although Christy (1890) knew of no definite record for Essex in later centuries.

Newton (1896) described it as 'formerly abundant in Britain'; this certainly seemed to be the case in the mediaeval period, judging by its frequent appearance in the lists upon which Table 1.3 (p. 20) is based and the numbers sometimes assembled for major feasts (e.g. Table 1.5, p. 22). However, during the 19th century, major drainage operations and, in places, increased hunting pressure across the bird's European range led to the abandonment of many breeding areas, and the bird was completely gone from the British Isles, most of Scandinavia and western Germany by the end of the century (Brusewitz 1969, Cramp & Simmons 1977, Kennedy *et al.* 1954, Hagemeijer & Blair 1997). Although Bitterns had stopped breeding in Britain by this time, they remained fairly common winter visitors, particularly in severe winters when large influxes of continental birds occurred. The numbers

Figure 4.1. *Historic breeding distribution of the Bittern in Britain by pre-1974 county. Revised from Shrubb 2003.*

of such winter visitors tended to decline in the later 19th and early 20th centuries, reflecting the further loss of breeding populations in northwestern Europe.

Figure 4.1 maps the historic breeding distribution of Bitterns in Britain by county, as far as it can be recovered from 19th- and early 20th-century county avifaunas. Lubbock (1845) observed that few nests were reported because of the treacherous nature of the birds' nesting sites in swamps and wet reed-beds, a point well-illustrated by photographs in Riviere (1930) and Percy (1951). Thus the distribution mapped rests largely on records of booming males, and may exaggerate the true nesting position. Stevenson (1870) also thought that the number actually breeding in Norfolk in the past may have been overestimated, owing to the number of migrant arrivals in late autumn. Sites given in the county avifaunas are often described in rather general terms, and some say no more than 'bogs' or 'major marshlands'. Nevertheless the avifaunas used to compile Figure 4.1 mention 70 sites where breeding birds or booming males were recorded, suggesting a reasonably healthy population up to the late 18th century.

Disappearance and re-establishment

As a breeding species in Britain, the Bittern had vanished from Wales and most Scottish counties by 1800, although it was gone from Sutherland in the 17th century (Baxter & Rintoul 1953) and finally disappeared from England in around 1860. This accords with the general pattern and timing of major land-drainage works in the 18th and 19th centuries (Shrubb 2003), which were undoubtedly the prime cause of this heron's decline. Not only were reed-beds totally lost but the lowering of water tables seriously degraded remaining areas of apparently suitable breeding habitat.

Bitterns re-established themselves in East Anglia in the early 20th century in a time of agricultural depression and deterioration of drainage systems. In Norfolk many marshland areas that had been traditionally mown and grazed were abandoned and became overgrown tangles (Riviere 1930). Taylor *et al.* (1999), for example, observed that the area known as Rush Hills on the south side of Hickling Broad, once a wader breeding site where Booth (1881–87) had obtained his specimens of breeding Ruffs in the 1870s, had become a reed-bed in 1920, where Jim Vincent found three Bittern nests (Taylor *et al.* 1999). Ennion (1949) gave a graphic picture of the rapidity of the reversion of farmland to fen in his book *Adventurers Fen*; the process took about 20 years from the point at which drains were neglected. Ennion found Bitterns nesting in Cambridgeshire in the 1930s, the first for 100 years in the county.

Hunting bitterns

Bitterns were always hunted, and before guns became widespread they were commonly taken in nets and with falcons, and probably with crossbows; snares and limed threads were also tried (Brusewitz 1969). But the early avifaunas provide little information on the methods used, and Darby (1974) noted that references to fowling in the Fens were also less frequent than might have been expected in mediaeval documents, with a dearth of material describing fully the regulations and customs that then governed it, although he had no doubt that these existed. Nevertheless, in the 13th century, royal officials and Fenland abbots and priors were charged with assembling and forwarding numerous wildfowl, including Bitterns, for royal feasts. These birds were presumably assembled and transported alive, so that they arrived fresh. Such activities continued into the 17th century; for example, Lilford (1895) quoted the Althorp Household Books for 1634, when 19 Bitterns were paid for in August.

In the 18th and 19th centuries, Bitterns were regarded as legitimate game by hunters and were widely shot for the pot. Lubbock (1845) recorded that in the late 18th century it was not unknown for parties of fen-shooters to kill 20–30 birds in a morning, and he shot 11 himself in 1819. Four to five could be seen in the Broads in a morning as late as the 1820s. Stevenson (1870) remarked that, when plentiful, they sold for a shilling each, the same price as Snipe; Alfred Newton (1896) was told by William Spencer, a thatcher of Feltwell, that his grandfather had one roasted every Sunday for dinner. At Whittlesey in Cambridgeshire, John Heathcote, in his diary, recorded one fenman shooting seven or eight in a morning in the early 19th century, and he and his brother once shot seven in one field of Holme Fen. Heathcote goes on to say 'I sank in the bog above my waist, and remained there till ropes and assistance were procured from a distant cottage to extricate me' (Lilford 1895, Wentworth-Day 1954).

In the mediaeval period, Bitterns' eggs were also collected and sold, presumably for food. For example, Darby (1974) recorded that in around 1300, the Littleport Rolls listed offenders who habitually collected Bitterns' eggs and sold them outside Littleport's Fen, for which they were fined. Macpherson (1897) recorded that it was once the custom for boys to wade out into the reed-beds to capture young Bitterns by hand, and Turner in 1544 (translated in Evans 1903) noted that Bitterns were driven into nets by means of stalking horses. Evans translated Turner's passage as

Stellaris is that kind which Englishmen denominate buttour or bittour and the Germans call pittour or rosdom. Now it is a bird like other herons in its state of body generally, living by hunting fishes on the bank of swamps and rivers, very sluggish and most stupid, so that it can very easily be driven into nets by the use of a stalking horse.

Brusewitz (1969) recorded that Bitterns were mainly hunted with bird-dogs but gives little detail.

Falconry

Salvin & Brodrick (1855) noted that Bitterns were once a favourite quarry at which to fly heron-hawks, but the practice had, I suspect, largely died out well before the 19th century; few of the 19th century avifaunas make any mention of it, and Stevenson (1870) remarked that he could find no record of it in Norfolk, and nor could Harting (1880 and 1886) nor Newton (1879). Nevertheless, Muirhead (1895) recorded that Bittern-hunting with falcons was a popular sport of Scottish kings in the 15th century. Brusewitz (1969) observed that taking them with falcons would have been fairly easy in summer, when they fly regularly between nests and feeding grounds. But opportunities for the kind of flights afforded by Grey Herons would have been fewer. However, Turner (1924) describes up to six Bitterns flying together in the air in May and June over Hickling, a time when the birds were numerous there. The birds circled round each other like gulls, frequently shooting up and then planing down again; this suggests that opportunities for falconry may have been more frequent when Bitterns were more abundant in the past (Figure 4.2). Ray (1678) noted that 'In the Autumn after sunset these birds are wont to soar aloft in the air with a spiral ascent so high till they are quite out of sight', a record that suggests birds departing on migration.

Figure 4.2. *Bitterns soaring over a breeding site in the 1920s (Alan Harris). This is based on the report of Turner (1924), who watched the Bitterns soar above Hickling Broad.*

The Night Heron

Prior to the 18th century, the Night Heron *Nycticorax nycticorax* was a widely scattered breeding bird in western Europe, as far north as Germany, and particularly common in the Low Countries (Yeatman 1976). Its population declined sharply during the 19th century, although a few pairs still bred in Germany before 1900 (Lippens & Wille 1969). In France it bred as far north as Strasbourg in 1849 but, before the Second World War, bred commonly only in the Dombes and the Camargue (Voisin 1991). Lippens & Wille noted that it ceased to breed in the Netherlands in 1876, although Lilford (1895) noted that adults and nestlings were sent annually from there to Leadenhall market in London at the end of the century, though they may not have been Dutch-bred birds.

In Britain, there is no evidence to suggest that there has ever been a wild population of Night Herons (Lippens & Wille 1969), despite Harting's (1901) observation that this species once nested annually, though feral populations have bred since 1951 (Thom 1986, Taylor *et al.* 1999). Bourne (1999a) quoted Thomas Muffett (in Mullens 1912) in support of the idea that the Night Heron, along with several other herons, bred in Britain in the mediaeval period. But Muffett was discussing their suitability as food and said nothing of breeding. Furthermore, although Night Herons appear in lists of birds consumed at royal feasts and in the price lists for the London market in the mediaeval period, they are not, unlike egrets, recorded in the contemporary household accounts I have examined. I believe they were always imported, as there was a significant import trade in such birds (for the 16th century see Bourne 1999b).

This species was one of the most keenly sought herons for the table, although I have found no account of its attributes as a table bird. It appears in English records as the Brewe, an anglicised version of the French name *bihoreau* (Bourne 1999a). Birds were harvested before they could fly and, as with other herons, fattened in cages or pens until required for the table.

Considerable numbers of Night Herons were harvested from colonies in the Low Countries. For example, Lippens & Wille (1969) recorded that 2,000 young birds were taken from a mixed heronry near Gouda in 1357. The same authors also describe a large mixed colony of Night Herons, Grey Herons and Spoonbills near Sluis (today on the Dutch/ Belgian border) in 1534; this was the property of the Abbey of Ter Doest, which reserved the right to harvest the young. From the 15th to the 18th centuries young Night Herons were also harvested annually from a site at Zevenhuiz, where Lippens & Wille (1969) suggest that a uniform harvest was taken annually throughout. Willughby and Ray visited this colony in the 17th century and were shown young birds, which they accurately described (Ray 1678). But they do not describe the taking of the young, although they do for Spoonbill.

Bourne (1999b) showed that at least 800 Night Herons were eaten on two State occasions held by England and France in the 1520s and 1530s. Such consumption led Lippens & Wille (1969) to consider that exploitation, coupled with the drainage of many of its breeding habitats, was among the main causes of the bird's rapid decline in Europe after the 18th century.

Egrets

Evidence from early texts, such as household accounts and bills of fare for feasts, show that egrets were well-known in mediaeval England until the 16th century, at least in the kitchen; they were also offered for sale by the London poulterers in the same period (see Tables 1.3, 1.4 and 1.5, pages 20–22). The name used in the mediaeval Latin of household accounts was *egres*, *egrecco* and variants, or *egerett*, without any qualifying adjective, which Woolgar (2006) translated as 'lesser white heron or egret' on the authority of the *Oxford English Dictionary*. Stubbs (1910a, b) also suggested that these birds were Little Egrets *Egretta garzetta* and that they had bred in England in the past; for evidence of breeding, Stubbs pointed to the Act for the Preservation of Grain of 1564, which stated that

> *this Acte shall not give liberty to any person to use any meane or engyn for the destruction of any Crowes Rookes Chawghes or other vermin to the Disturbance Lett or Destruccyon of the building or breeding of any kinde of Hawkes, Herons, Egrytes, Paupers, Swannes or Shovelers …*

One should be cautious in accepting such legislation as evidence of contemporary breeding, as such regulations may have persisted long after the species ceased to breed (see Smith 1887, p. 391).

There are, in fact, very few historic records to support the notion that breeding by any egret species took place in Britain. The only record suggesting a breeding colony that I have found is in the household accounts of the Bishop of Salisbury in the early 15th century, edited by Woolgar (2006). In these accounts Ralph of the Buttery twice went to Wimbourne to buy herons and egrets and, on the second occasion, Spoonbills. On his first visit eight egrets and 20 herons were bought, and Ralph was also paid for the expense of carrying them back to Salisbury, including food for the birds, indicating that they were transported alive. On the second visit, in May 1407, Ralph is recorded as simply going for herons, Spoonbills and egrets, which were acquired for the funeral feast of the Bishop (Woolgar 1999). It is a reasonable assumption that these birds were taken from the nest before they could fly, which was the usual way of harvesting herons. These records suggest that there was a mixed colony of heron species – including egrets – at Wimbourne at this date. It is of particular interest that Spoonbills were involved, as all the sites for which any detail is known for this species in England in the past were in mixed colonies with Grey Herons (see p. 63).

One other record of interest in Woolgar (2006) concerns the accounts of Eleanor of Brittany, an unfortunate relative of King John who spent most of her life in captivity. The accounts record the purchase of 17 egrets in August, September and October 1225, at which time Eleanor was captive at Bristol Castle. They were accounted for at the rate of one or two (once three) per week, with three weeks blank. This pattern strongly suggests that the birds were held in stews, and drawn on as required. Their original source, however, is not indicated.

As with the Night Heron there is no need to assume a local breeding population to account for the appearance of egrets on English tables, as they were undoubtedly imported (Bourne 1999b). Bourne also noted that English importers had to go further south in Europe to secure egrets in the 16th century, perhaps as the result of over-exploitation. If Willughby and Ray met with breeding Little Egrets on their European travels, they did not report them, and their description of the bird was taken from a specimen purchased in Venice market (Ray 1678), which hints, perhaps, that it was already a southern species in their day. But for the Little Egret, cold weather in winter is a major cause of mortality in Europe (Voisin 1991). The sharp deterioration in the climate of northwest Europe from the mid-14th century known as the Little Ice Age, which was at its most severe from the later 16th to the early 18th centuries (Fagan 2000), probably contributed to their decline and range contraction. Little Egrets disappeared from much of Europe in the late 19th and early 20th centuries, the result not of climate change but of the depredations of the plumage trade. It must be the case that its present range expansion into former breeding territories is as much the result of protection as a change in the climate.

The Grey Heron

This is much the most numerous heron species in Britain and in continental Europe, one that has increased considerably since 1970 (Birdlife International 2004). In the past, Grey Heronries were valuable properties, protected to supply meat for the estate owner's table, usually in the form of young birds, and to provide sport for falconry. The practice of taking a regular harvest of young birds for consumption persisted well into the 17th century when, as in France, English heronries were harvested annually for young birds for a good profit (Ray 1678). Some estates founded heronries on their land in the 14th century. This seems implicit in Harvey's (1991) observation that one was started on the manor of the Argentein family at Wymondley, Hertfordshire, in the 1350s. Stone (2006) made a similar observation of a heronry at Hinderclay in Suffolk, which produced a significant income for the manor from the sale of young birds. Bailey (2007) recorded that heron farms were also developed on manors in the Waveney valley in Suffolk in the 15th century. One at Mettingham produced around 40 herons a year, with some for sale. This was part of the trend noted in Chapter 1 (see p. 21), towards the increased production and consumption of birds in the aftermath of the Black Death. But the 'planting' of heronries was still being carried out on occasions, until as late as the early 19th century. Thus Knox (1849) recorded that the extinct heronry at Penshurst in Kent was founded in the reign of James VI/I by birds brought from Coity (in Glamorgan) by Lord Leicester's steward; Oakes (1953) recorded that a heronry in Hamilton Wood in Lancashire was founded by the Duke of Hamilton, who brought herons from Hamilton in Scotland and liberated them at Ashton Hall between 1800 and 1810. The Penshurst colony died out in the early 19th century, although the trees that supported the birds' nests were still standing in the early 20th century (Ticehurst 1909). By contrast, the Ashton Hall colony is still occupied, with five pairs in 2003.

Although Nicholson (1929) recorded that in the 1830s Lord Carnarvon's friends at Pixton, Somerset, ate Grey Herons skinned, stuffed and roasted like hare, with strawberries and cream to follow, Pennant (1776) noted that the species was falling rapidly out of favour as a table bird in his time. Eating herons was not confined to the upper classes, and country people killed or trapped full-grown birds at all seasons until the end of the 19th century; Alexander (1896) noted that young Grey Herons were still being shot, largely for eating, on Romney Marsh in May–June.

Heronry demographics

Little detailed information on Grey Heron numbers in Britain is available before the end of the 19th century, but some interesting details about heronries are evident. Since the mid-19th century, the size of heronries in England and Wales has declined. This pattern was indicated clearly by Nicholson (1929), who gathered counts totaling 2,040 nests from before the mid-1860s for 32 heronries, an average of 64 nests per colony. For the second half of the 19th century, mainly after 1872, Nicholson had counts for 106 heronries, which averaged 32 nests per colony. In the 1928 and 1954 national heronry surveys, average colony size was between 14 and 15 nests, and in the most recent national survey in 2003 it was between 15 and 16 (Marchant 2004). Not being single-year surveys, the 19th-century counts are not strictly comparable with those of the 20th century. But the difference in colony size is marked enough that it is difficult to believe that this is not a genuine trend.

That this trend has not occurred against a background of a similar decline in overall population (see Marchant 2004) indicates that many old colonies have broken up and shifted. In Scotland Baxter & Rintoul (1953) also observed a definite tendency for the big heronries to break up and the birds to nest in smaller units, with them considering that this was associated with an overall decline in population. But in north and west Scotland the species was expanding and simultaneously dispersing into more numerous but smaller colonies up to 1940 (Marquiss 1989).

The shifting distribution of heronries in England and Wales is clear from examining Nicholoson's (1929) results. He found that only 55% of the colonies visited for the 1928 national survey were active in the second half of the 19th century. Of these he classified the age of 78 as 'immemorial' – defined as a heronry that had existed as long as anyone questioned could remember, but with no traceable evidence or tradition of its origin – and 22 as 'ancient', with evidence of their existence pre-dating 1800; 50 and 18 respectively of these sites were still active in 1954 (Burton 1956). Their status today is listed in Appendix 1.

The pattern of dispersal into more and smaller units was originally considered to have derived from the change in the status of the species in the 19th century, from one valued and protected for food and for sport – in effect a gamebird – to active persecution, with the destruction of colonies and birds because of perceived damage to fisheries, as opposed to managed exploitation. This explanation does not seem entirely satisfactory. Because of their distribution in the parks and estates of the landed gentry most heronries still enjoyed active protection in Nicholson's day, and he remarked on the stability of the total population (Nicholson 1929). He had records of fewer than 20 colonies that had been shot out, mainly before 1914. But herons were also at risk away from colonies, and Nicholson listed Derbyshire, parts of Devon, Gloucestershire, Hampshire, Herefordshire, Leicestershire, parts of Lincolnshire, Nottinghamshire, Westmorland, Wiltshire, parts of Yorkshire, Anglesey and Caernarvon as counties in which large numbers of Grey Herons were killed, mostly by fishing interests. The impact of shooting herons on fisheries away from colonies could be severe. For example, it reduced the ancient and strictly protected colony at Althorp from 100 nests in 1842 to an average of 10 in the 1880s (Lilford 1895). But it is not clear why this persecution should have caused colonies to break up and shift.

Voisin (1991), however, drew attention to the marked difference in colony size between continental Europe and Britain, observing that the size of heronries is proportional to the richness of foraging areas within a radius of *c.*25km and noting the much more extensive wetlands available to Grey Herons in Europe. The decline in heronry size in Britain seems to coincide quite well with the great surge in land drainage in the 19th century. This not only affected many major wetlands, particularly between about 1780 and 1850, but, in the period up to the mid-1870s, also involved a massive expansion of field drainage, which undoubtedly affected water tables (Shrubb 2003). There was significant deterioration in drainage after 1875, particularly between 1920 and 1939, but another surge in drainage activity started with the Second World War and has continued to this day (Shrubb 2003). This must have reduced the extent and probably richness of many feeding areas, although this has apparently been insufficient to reduce the overall population of Grey Herons.

Falconry

Nicholson (1929) suggested that one once-considerable toll upon the species was the sport of falconry. He commented that even when the bag was insignificant, the disturbance to the heronries involved and the ensuing mortality among the young must have been extensive. At Althorp, Lilford (1895) recorded the construction in 1603 of a lodge with an open gallery around the first floor to enable ladies and gentlemen to watch the heron-hawking. The practice of heron-hawking continued into the 19th century, but the number of practitioners in England declined rapidly with enclosure, which restricted the unimpeded access to open country that the falconers needed; heron-hawking died out in Britain in the mid-19th century.

Probably the best-known English heronry for falconry in the latter years of the sport was that near Brandon, at Didlington in Norfolk. This heronry was certainly in existence in the 1770s, and falconry was practiced there until at least the mid-1840s (Salvin & Brodrick 1855). The earliest nest-count seems to be for 1868, when there were 60–70 nests. The colony died out in 1924, long after falconry had ceased, but it was reestablished in 1953 (Nicholson 1929, Taylor *et al.* 1999).

I doubt Nicholson's view of the impact of falconry on herons. The sport of heron-hawking depended on certain conditions to be effective. It needed a substantial colony and rich feeding areas, well-separated by open country over which falconers could ride unimpeded in pursuit of their birds. By no means all heronries were suitable. The most famous 19th century European site for this sport was at the old Royal palace of Het Loo in the Netherlands, where there was an ancient heronry that contained around 1,000 pairs, separated from their feeding areas by two extensive areas of heath (Van der Wall 2004).

Figure 4.3. *A 16th-century falconer with a cadge carrying six falcons. Peregrines and Lanners, which these birds appear to be, were the falcons most favoured for heron-hawking. Below the figure is a copper band, which was used for marking the herons taken. From Harting 1971 and Salvin & Brodrick 1855, the latter courtesy of Beech Publishing House.*

The object of the sport was not to kill herons but to enjoy the spectacle of the flight. The heron-hawking season was in the spring, when young were in the nest and the falcons were flown at birds returning to the colony with food for the young. Falcons were flown in pairs. On seeing the falcons in pursuit the herons started to climb, with the falcons behaving similarly in an attempt to get above to stoop. Eventually after several stoops one falcon would bind on to the heron and bring it down, releasing it just before hitting the ground. The falconer galloped up, gathered the falcon and secured the heron. After examining the heron to see it was unharmed, it was marked with a copper band on the leg with the date and place of the flight and released (Plate 2 and Figure 4.3). It was not unusual to find herons carrying two or more such rings (Salvin & Brodrick 1855). Harting (1880) quotes two instances in Europe where birds so ringed were recovered, presumably breeding, 10 or more years later.

Whilst this sport may have affected breeding success, it did not disrupt the colonies concerned. One reason was that the action tended to take place well away from the colonies. The heaths over which the sport took place at Het Loo, for example, were 5km from the heronry itself; nor were feeding areas disturbed. Furthermore, herons were infrequently killed, despite the number of herons taken; falcons that habitually damaged their herons were taken out of use (Van der Wall 2004). Van der Wall showed that between 1839 and 1852, around 2,300 herons were taken at Het Loo. Most were examined, marked and released, but a few were taken into captivity for use in training new falcons. This type of falconry was not developed much before the Renaissance and, if hawks were used for taking herons for the table in the mediaeval period, Goshawks *Accipiter gentilis* may often have been used (Salvin & Brodrick 1855).

Taking herons for the pot

Herons were also taken with the longbow and crossbow, and legislation was introduced in an attempt to restrict the use of the crossbow (see p. 41), presumably as it was too efficient a weapon. In addition, herons were sometimes caught on hooks baited with small fish (Macpherson 1897). But the main method of acquiring herons for the table was undoubtedly taking young birds before they could fly. Most country estates that owned a heronry kept equipment for climbing to the nests and hooking down young from the nests or branches. In the account books of Herstmonceux Castle for 1643–49 there are references to payments for climbing for 162 herons, making a new heron rope, for white leather for the heron climber's use, and for a pole for his heron-hook (Lennard 1905), and similar payments are referred to in other early household accounts, such as those of the Le Stranges at Hunstanton (Gurney 1921) and the Spencers at Althorp (Lilford 1895). At Althorp, Lilford noted payments for climbing for herons in late April and late June, suggesting that second or repeat broods were exploited. Young herons were regarded as worthy presents between landowners, and payments of gratuities to men delivering such gifts also feature in household accounts. The birds were rarely used at once, but were instead kept in pens (or stews), and fattened on butchers offal, curds and similar foods. Household accounts often refer to such items and to the construction of the pens. At Althorp the birds were kept in sheds, and there were payments for cleaning these in November (Lilford 1895). Lilford also noted that the Spencers at Althorp employed a 'cram mayde' to feed up poultry and herons. Her duties presumably included force-feeding the birds, probably with oatmeal.

Considerable numbers of young herons were taken and used in this way. The great feasts included in Table 1.3 (see p. 20) involved a total of 1,570 Grey Herons served at 13 such events. The heronry near Gouda from which Lippens & Wille (1969) recorded the harvesting of 2,000 Night Herons in 1357 also yielded 564 Grey Herons in the same year. The Herstmonceux records referred to above involved the heronry now at Wartling in Sussex. It held 150 pairs in the 16th century and 117 in 1872 (Harting 1872). It seems unlikely the culling of 162 young would have had a great effect on a colony that has persisted for 400 years. The high value placed on herons for sport and food meant that their exploitation was accurately and carefully managed.

The Spoonbill

The Spoonbill *Platalea leucorodia* is predominantly a bird of southeastern Europe and Asia. Its range in the western Palearctic has always been scattered, and its distribution in western Europe has been much reduced by drainage and the exploitation of eggs and young (Cramp & Simmons 1979). In the past in western Europe it was known to breed in the Low Countries, in England and Wales until the 1660s (Newton 1896), on the borders of Brittany and Poitou in France in the mid-16th century (Newton 1896), in Italy near Ravenna in 1500 (Cramp & Simmons 1979), and near Santarem, Portugal, in the early 17th century (Tait 1924). Newton (1896) noted that by the late 19th century its northwest European breeding population was restricted to the Netherlands.

In England and Wales, Spoonbills bred or probably bred in seven sites in Norfolk, namely Hunstanton, Wormegay, Fincham, Poppylot, Whinburgh, Cantley and Reedham (Gurney 1921); Sir Thomas Browne also gave Claxton, which is on the opposite side of the Yare to Cantley and may be the same site (Harting 1877). In Suffolk it bred at Trimley about 1668, which was apparently the last English occupied site. Spoonbills also bred at Fulham in the Bishop of London's park (Harting 1901) in the early 16th century, in two woods in East Dean park in Sussex in 1570 (Harting 1877), in Cobham Park in Kent in the 16th century (Ticehurst 1909) and in Pembrokeshire in the 16th century (Mathew 1895). There was probably also a colony at Wimbourne in Dorset in the early 15th century, and the accounts of the Earl of Oxford, in Essex during 1431–32, record receiving young Grey Herons, Spoonbills (*Popelles*, presumably young birds) and Bitterns from his manor at Great Bentley (Woolgar 2006), indicating the presence of a mixed colony of Grey Herons and Spoonbills there. Lorand & Atkin (1989) also found a record for Lincolnshire in Dugdale's 1662 *History of the Drainage of the Fens*, in which a map dated 1662 marks a wood near Crowland as 'Dousdale Holt where the white herons breed'. They noted that William Charleton in his 1668 publication *Onomasticon Zoicon* listed 'White Heron' as an alternative name for Spoonbill, although early alternative names for this species were usually Shoveler, Shovelard or Popeler. Nearly all the sites for which any detail is known saw the birds sharing mixed colonies with Grey Herons in trees.

It is also relevant here that Robert Laneham, describing a visit by Queen Elizabeth I to Kenilworth in Warwickshire, noted that on the bridge by the castle 'upon the first pair of posts were set two comely square wire cages, three feet long and two feet wide, and high in them, live bitterns, curlews, shovelers [spoonbills], hernshaws [herons], godwits and such like dainty birds …' Whether these 'shovelers' were locally taken cannot be known, but the record, quoted by Lubbock (1845), provides an interesting insight into how such species were handled.

Disappearance from Britain

There is little doubt that over-exploitation, perhaps in an era of unfavourable climatic conditions, the Little Ice Age, was mainly responsible for the disappearance of this species as a breeding bird in England and Wales. The latest record we have is from Sir Thomas Browne, writing in *c*.1668, who noted that Spoonbills 'have formerly bred in the Hernery at Claxton and Reedham; now at Trimley, in Suffolk' (Harting 1877), a record that inferred that, by that time, the sites in the Yare valley at Claxton/Cantley and Reedham were no more. Gurney's (1921) account leaves the impression that few of the East Anglian sites were active much after the 16th century. Browne went on to say 'they come in March, and are shot by fowlers, not for their meat, but the handsomeness of the same; remarkable in their white colour, copped [crested] crown, and spoon or spatule-like bill' (Harting 1877) – an early instance of persecution for plumage.

Browne's report apart, there is little evidence that adult Spoonbills were taken in Britain. Virtually all the records I have seen have referred to the taking of young before they could fly, although Macpherson (1897) recorded that Dutch fowlers concealed themselves in the reeds at breeding sites and captured adults with a hand-net, wielded like a butterfly-net. Willughby and Ray visited the large mixed heronry at Zevenhuiz and watched the young being taken, saying 'when the young ones are ripe, those that farm the Grove with a hook on the top of a long pole catch hold of the bough on which the nest is built and shake out the young ones, but sometimes nest and all [fall] down to the ground' (Ray 1678). Interestingly, Pennant (1776) noted that this colony was extinct in his day, the trees having been felled.

Spoonbills were also exploited extensively for their eggs, despite Cott's (1953) description of them as scarcely fit for consumption. The literature has little to say about this in Britain, but that it was a subject of concern is indicated by the Act of 1534 (see p. 45) which, among other species, specifically protected the eggs of Spoonbills on a penalty of 8d for each egg taken, legislation that Newton (1896) described as futile in the absence of any protection for the breeding adults. Sclater & Forbes (1877) remarked that Spoonbills' eggs, imported from Holland, could be purchased cheaply in London egg-shops, and gave a detailed description of the farming of the colony at Horster Meer, between Amsterdam and Utrecht. The eggs (and those of Cormorants) were collected from the nests twice-weekly in May and June, and sold in Amsterdam. Collecting ceased 'at the period when it is known by experience that the birds cannot lay more eggs', at which time the birds were allowed to complete their nesting cycle. The Meer was owned by a rich proprietor in Amsterdam, and was farmed out 'at a considerable rent' for the sake of the fish, reeds and birds' eggs it produced.

The tradition of farming Spoonbills as a food source thus had a long history in the Netherlands. To what extent demand from England could be met from English sources is unclear. But the limited number of breeding sites that we know of suggests that demand had to have been met by importing birds from the Netherlands, a trade confirmed by William Harrison in 1577 (in Gurney 1921).

The Crane

Two species of cranes breed in Europe, of which the Common Crane *Grus grus* is the most numerous and is the species discussed here. It once bred over much of Europe. Bones found during archaeological excavations have shown that it was widely distributed in Britain from prehistoric times (Boisseau & Yalden 1998), although such records give no information on status or abundance. It probably bred in Ireland until the 11th century (Armstrong 1979) and perhaps until the 14th century (Gurney 1921), and bred in England and Wales until the end of the 16th. Breeding was perhaps last recorded in East Anglia in 1590 (Saunders 1899). There is no certain breeding record for Scotland. But the Scottish records given by Gurney (1921) include seven birds for April and four for May (to the 30th), which fall well within the present nesting season in Scandinavia given by Cramp & Simmons (1979).

How abundant a breeding bird the Crane was in Britain is not known, but from the later mediaeval period, at least, the species was probably always more numerous as a winter visitor and passage migrant. Figure 4.4 illustrates the months in which Cranes were provided for feasts or as presents to notable personages from the 13th century. The bulk of the spring records were for Scotland, some of which may have been breeding. Otherwise only 7% of the records were for June and July, when birds almost certainly were breeding, while 85% were for autumn and winter. These records should be regarded with some caution as they may not be fully representative, with seasons for many similar records unstated. Cranes may also have been easier to obtain in winter, especially with falcons, when they were in fields rather than the more inaccessible marshes used for breeding. Furthermore, some (at least) of these birds may have been supplied from captive or reared stock.

The species ceased appearing regularly after the 17th century. Pennant (1776) noted that in Lincolnshire and Cambridgeshire, where Ray (1678) noted large winter flocks in his time, the local inhabitants were scarcely acquainted with them by the mid-18th century.

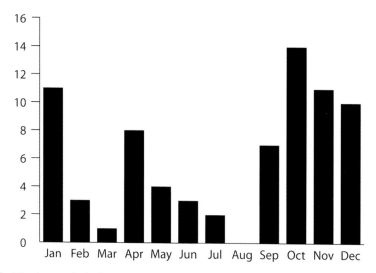

Figure 4.4. *Months in which Cranes were provided for feasts or as presents for distinguished guests in Britain. Data from Southwell (1901), Nelson (1907), Gurney (1921), Stubbs (1910a), Armstrong 1979, Bourne (1981, 1999b), Rackham (1986), Lorand & Atkin (1989).*

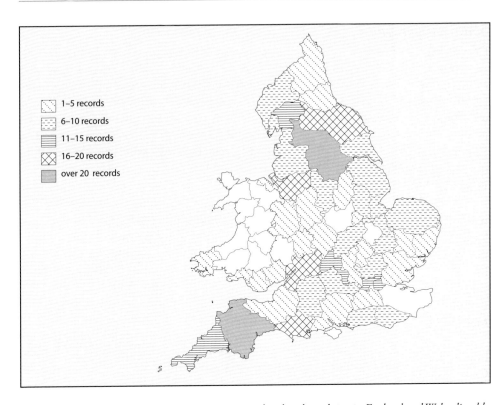

Figure 4.5. *The distribution of Crane place-names and archaeological sites in England and Wales, listed by Boisseau & Yalden (1998) by pre-1974 county.*

He concluded that Cranes were no longer regular visitors to Britain, and Naylor (1996) could only list two records for the late 18th century, for 1773 and 1797, and only 14 more until the 1860s. Nevertheless Thirsk (2007) quoted a cookery book written by Sarah Jackson that offered recipes for Cranes in 1754, so they may still have been obtainable at that time.

An indication of the past distribution of Cranes in England and Wales can be derived from the frequency of place-names referring to them. These were analysed from the publications of the English Place Names Society (EPNS) by Boisseau & Yalden (1998), who found 271 place names that included references to Cranes, mainly in England. Some records were found in most of the counties that have been surveyed by EPNS, and at least half this sample had the name associated with wetland features. Figure 4.5 maps the distribution of Crane place names and archaeological sites listed by Boisseau & Yalden in England and Wales by pre–1974 county, to illustrate the general nature of the species' early distribution. As with archaeological evidence, place-names give little indication of whether Cranes were migrants, winter visitors or breeding birds; they indicate only that the bird was familiar and widespread.

Other evidence of the Crane's familiarity can be found by observing the frequency with which it was illustrated in the illuminations of mediaeval manuscripts. Yapp (1981) noted the Crane as the bird illustrated most commonly in English manuscripts, with illustrations going back to the 11th century and occasionally including hawking scenes. Cranes appear just as frequently in continental manuscripts, but Yapp observed that he found no well-drawn examples there from prior to the 15th century.

Status in Scotland

Documentation of Cranes in Scotland is much sparser. Yalden & Albarella (2009) list only six archaeological records, between the Bronze Age and the early mediaeval period, with all except one (at Newstead in the Scottish Borders) at coastal or island sites. Armstrong (1979) commented that one reason for the sparseness of the Scottish archaeological record was that most excavations had been done in rocky coastal sites. The first documented record is apparently for 1503 in Dumfries (Bourne 2007) and, from the household accounts of King James V between 1525 and 1533, Bourne listed records from Coldstream (Borders, three records), Edinburgh (Lothian, five), Fife (seven from four sites) and Stirling (two). The Bishop of Aberdeen is also recorded sending presents of Cranes to the King in these accounts.

Cupar Abbey in Fife obtained Cranes from a fowler in the 1540s and 1550s. Although there is no certain breeding record for Scotland, Armstrong quoted John Leslie, Bishop of Ross, writing in 1578, that the Crane then inhabited Ross and Inverness, an observation that suggests residence. Similarly, Martin (1703) remarked of Skye that there were 'plenty of land and water fowl in this Isle as hawks, eagles … blackcock, heath-hen, plovers, pigeons, wild geese, Ptarmigan and Cranes'. He recorded a flock of 60 Cranes on a beach there in the late 17th century. Cranes became extinct in Scotland in the 18th century (Darling & Morton-Boyd 1964).

There are few definite reports of breeding localities in Britain, at least partly because by the time ornithological records began to be listed and published in the 16th century, Cranes were already declining fast as breeding birds. Some evidence of this is provided by legislation protecting the eggs of Cranes and other species in 1534 (see p. 45), a measure suggesting increasing scarcity, and designed to preserve Cranes for the King's use rather than any idea of species conservation. George Owen, in his *Description of Pembrokeshire* published in 1603, wrote that Cranes bred in the bogs of that county, and clearly distinguished them from herons, which he noted bred in trees (Mathew 1895). Dr William Turner, in his *Avium Historia* of 1544, recorded that he frequently saw young (known as pipers) (Evans 1903). It has often been assumed that Turner made these observations while living in Cambridge, and that they refer to the Cambridgeshire Fens. But Turner's statement that 'among the English also, Cranes nest in marshy places, and I have very often seen their pipers' (Evans 1903) clearly seems to imply a wider distribution. Southwell (1901) also recorded young taken at Hickling Norfolk in June 1543. Moreover, the dates Cranes were supplied as gifts or for feasts (Figure 4.4, p. 65) sometimes indicate their presence in the breeding season in counties such as Lincolnshire, Norfolk and Nottinghamshire, and perhaps in Scotland (see references in Figure 4.4).

Domestication

Cranes are readily domesticated. Yarrell & Saunders (1884) noted that, when taken young, they made amusing if somewhat dangerous pets. Some international traffic in cranes existed from at least the 15th century, when Crowned Cranes were being kept in Rome (Johnsgard 1983). Keeping Cranes in captivity as pets or to fatten up for the pot is a practice that goes back to ancient Egypt. They were commonly illustrated in Egyptian art, usually as domesticated or captive birds, being herded, fed on grain, occasionally being force-fed (Figure 4.6) and sometimes shown hobbled by having their bills tied to their necks

Figure 4.6. *Fattening cranes in ancient Egypt. From the Tomb of Ti, in Moreau (1930).*

preventing them taking off (Houlihan & Goodman 1986). Houlihan & Goodman could only trace two illustrations, which were of birds being taken in clap-nets, of Cranes that were not domesticated. They considered it most likely that the Cranes used to stock Egyptian aviaries were trapped on migration, although some may have been bred in captivity. Cranes were also domesticated in ancient Greece as pets and for the pot and, according to Plutarch, even nested as tame birds (Armstrong 1979).

Cranes were one of the species that Cistercian monks were forbidden to keep as pets in the 12th century, and were also kept in captivity in compounds by the Holy Roman Emperor Frederick II in the mid-13th century, so they were always available for his falcons and presumably for his table (Yapp 1982). Yapp remarked that there were no similar English records, but Edward Topsell, writing in the early 17th century, also spoke of tame Cranes (Harrison & Hoeniger 1972), and his account leaves the impression that keeping and breeding these birds was not particularly uncommon in the past. Topsell also recorded that Cranes were fattened in darkened rooms. Harting (1882) quoted an English inventory of 1500 that listed three live Cranes, valued at five shillings, implying that they were held in captivity. Stubbs (1910a) recorded that, for King Richard II's feast in 1387, the cook had to kill the Cranes provided, again suggesting that they were held in captivity, although not necessarily for any length of time. It certainly seems probable that the logistics of assembling the numbers of Cranes quoted for some of the mediaeval feasts rested at least partly on the availability of captive supplies, as did the accumulation of large numbers of various herons for the same events.

Hunting and falconry

Cranes were taken with crossbows and, from the mid 16th century, with guns, and the increased effectiveness and availability of shotguns probably contributed to their demise in Britain. Certainly Brusewitz (1969) considered that they were among the birds worst affected by the spread of such weapons. Cranes were also snared and taken with bird lime, although some of the methods described for the latter seem improbable. Thus Macpherson (1897) related that Cranes were said to be taken by the ancient Greeks by baiting a limed gourd with a live insect. When the Crane heard and tried to take the insect it was blinded by 'its sticky nightcap' and could be captured alive. Edward Topsell (in Harrison & Hoeniger 1972) also tells this tale, and Brusewitz (1969) illustrated a 16th-century European engraving of a variation of this trick, using limed cones thrust into holes in the ground; hollowed pumpkins were also supposedly used, as were limed strings. How accurate these stories are seems uncertain; their being recorded in early hunting manuals is not necessarily proof of credibility.

Accounts of how snares were used are more convincing, with snares or series of snares set at baited sites so that the birds' feet or necks were entangled, a common theme with many birds. Macpherson (1897) records eyewitness accounts of such methods used to catch Cranes in India, and essentially the same approach was used in Germany (Brusewitz 1969). It is of interest that there seems to be no account of such methods being used in Britain.

Above all, Cranes were favoured quarry for the falcons of kings, queens and aristocrats. The falcon most often used was the Gyrfalcon, a name that, according to Johnsgard (1983), derives from the German for 'Crane Falcon'. But Harrison & Hoeniger (1972) suggest an alternative derivation, from the latin *gyro*, to circle, from the Gyr's habit of circling its prey before swooping. There was a considerable international trade in Gyrfalcons from the early mediaeval period (Potapov & Sale 2005). In eastern Europe the Saker was also used (Salvin & Brodrick 1855), and Topsell (in Harrison & Hoeniger 1972) noted that small eagles were sometimes used in Germany, and that smaller falcons were used there to force Cranes to land, so that they could be caught by greyhounds. According to Thomas Muffett, writing in the late 16th century, the young were also taken with Goshawks (Mullins 1912).

We have little exact information about the scale of such hunting, but Gurney (1921) observed that Cranes were continually mentioned in connection with hawking and were the favourite quarry of mediaeval falconers if they could find them. Both Welsh and English princes gave significant rewards to their falconers for taking them, and Gurney noted that King John issued safe-conducts (licences) to fowlers to go to various parts of the kingdom to catch Cranes, for supplying the palace when he was not hunting.

Cade (1982), in his description of the hunting behaviour of the Gyrfalcon, provides a clear explanation of why they were favoured for hunting birds such as Cranes. He noted that Gyrs catch most quarry only after a long chase, which can sometimes go on for several kilometres, and that they are excellent at chasing quarry up into the air and forcing it to remain aloft until exhausted, at which point it will attempt to dive for cover. In this way Gyrfalcons catch cranes and other birds with light wing-loadings and superior powers of rising rapidly in spirals. The Gyr, however, does not 'ring' but climbs straight up at a steep angle, often making only one turn before it has gained ascendancy over its intended prey. This climbing ability made the Gyrfalcon the favourite hunting hawk in times past for flights directly off the falconer's wrist at large, high-flying quarry such as herons, cranes and kites. The essence of the sport of falconry was the flight; this description suggests how dramatic such flights must often have been.

Extraordinarily, there seems to be no clear indication that Cranes were ever caught during the flightless period of their moult, which occurs in June and July when the young are also still flightless. They would have been at their most vulnerable at this time, and this would have been the most logical period to acquire captive birds. Gurney (1921) does in fact suggest that young may have been captured in the Fens and sold in Cambridge market. Mediaeval Englishmen had great experience of trapping large numbers of wildfowl at the same stage of moult (see p. 76). It is hard to believe that no attempt was made to trap Cranes in the same way, but no record appears to have been left.

Although a highly regarded gamebird in Germany and eastern Europe, Cranes were actively controlled because of the damage they were believed to do to crops. In Poland, farmers built watch-huts in their fields to guard their crops, and the birds were hunted with stalking horses and long-range shotguns, often nine-barrelled carriage guns (Brusewitz 1969, Hobusch 1980). Even today, wintering birds are still scared away from young cereals in northern Israel by lines of men walking across the fields banging dustbin lids and firing guns (pers. obs.).

Cranes (and storks) were systematically and legally persecuted in Germany until well into the 19th century. In Mecklenburg they were declared fair game, with no close season, until 1934. By a royal decree of October 3rd, 1722 everybody in Prussia was allowed to shoot or catch Cranes, although care had to be taken not to confuse them with bustards. There was little or no protection anywhere on their migration routes. Where close seasons existed they were fixed for May and June (Hobusch 1980), ineffective to protect breeding birds, as breeding starts in early April and continues into July and August (Cramp & Simmons 1979). In the 19th century stocks were dwindling rapidly everywhere as the result of such persecution, combined with drainage of the birds' wetland haunts.

Long-term decline

As breeding birds, Cranes disappeared early from western Europe. The long-term trend from the 16th century seems to have been a general tendency to retreat to the north and east. Although the bird still bred in Spain in the late 19th and early 20th centuries, the Crane was much more abundant in Spain both as a breeding bird and a winter visitor in the mediaeval period (Bernis 1960). It had ceased to breed in Britain before the major drainage schemes of the 17th century affected their fenland breeding habitats. Indeed, the only area of western Europe where large-scale drainage and marked intensification of agriculture occurred earlier was, perhaps, the Low Countries, where such changes were in progress from the 14th century; large areas of waste were being brought into cultivation in Friesland and Gronigen early in the 16th century (Israel 1995). But it is not clear that Cranes ever bred there. It seems likely, therefore, that the early loss of Cranes in the west resulted from over-exploitation and persecution more than from habitat change.

Declining breeding populations in more westerly areas must also have resulted in declines in winter visitors and passage migrants there. This was certainly so in Spain (Bernis 1960) and is clear from Armstrong's (1979) quotation of the Italian F. Redi who, writing in 1671, reported great concourses of Cranes on the plain at Pisa, where no memory of such gatherings now remains. Willughby and Ray recorded seeing many Cranes offered for sale in Rome around the same time (Ray 1678). Nevertheless, the strongest argument for persecution being a major driver of this species' decline in the past is its population recovery today, in an era of great agricultural intensification but total protection (Hagemeijer & Blair 1997; Parkin & Knox 2010). This has seen Cranes reoccupying breeding areas last used in the 16th century, including the Fens.

CHAPTER 5
Wildfowl

Wildfowl – the swans, geese and ducks – were, and possibly still are, the most important group of wild birds to humanity, providing food, feathers and, latterly, sport. Archaeological research has shown that wildfowl have always been exploited (see p. 17), for they have particular advantages for hunters, being abundant, meaty, usually good eating, having copious plumage and down and often occurring in considerable concentrations.

Ducks and geese were also domesticated early – geese, for example, in Homeric Greece, whilst ducks were kept in substantial aviaries on Roman estates in the first century BC (Toynbee 1973). Most existing breeds of domestic duck are descended from the Mallard *Anas platyrhynchos*, and of geese from the eastern race of the Greylag Goose *Anser anser* and the Swan Goose *Anser cygnoides*. But there is little doubt that native populations of Greylag in Britain were also domesticated. Pennant (1776) recorded that the young were often taken and easily made tame, and 'were esteemed most excellent meat'. Their eggs were also taken to hatch and the young reared for the pot.

The Muscovy Duck *Cairina moschata* of Central and South America (the name, like that of the Turkey *Meleagris gallopavo*, is a complete misnomer) was also domesticated, and was probably brought to Europe by the Spanish. It has been widely introduced elsewhere, being recorded, for example, as a familiar sight in West African villages (Landsborough Thompson 1964, Kear 1990). Kear listed the favourable attributes of wildfowl that would have attracted farmers as longevity, good parenting, good food conversion rates, rapid growth and maximum production of down and feathers.

Until the development of reasonably efficient flintlock sporting guns, the main means of taking wildfowl were nets, duck decoys, snares and occasionally bird lime. They were also shot by bow and by crossbow, and taken by falconers. Nets have been used since prehistoric times. Much netting was done during the flightless period of the moult and before the young could fly, which provided opportunities to obtain large supplies quickly. Duck decoys, which evolved from such activities, are considered in a separate section (see p. 78). In this chapter I examine the exploitation of each of the three groups of wildfowl.

Swans

It is only since the early 20th century that most of the swans have had some measure of protection. Previously they were harried in summer and in winter, for food, trade and sport (Dawnay 1972). Dawnay remarked that the 'sporting chronicles of the 18th and 19th centuries are riddled with dead swans'. Hawker's (1893) Diaries contain several examples of this, and record that, in total, he shot 38 Whooper Swans *Cygnus cygnus* on the Hampshire coast between 1802 and 1853. Similarly, the the sight of a swan at any time in 19th-century America was the signal for every man with a gun to pursue it (Kortright 1942). Kortright noted that some species were more vulnerable than others because of behaviour and distribution. Thus the habit of the Trumpeter Swan *Cygnus buccinator* migrating in small parties along the coastline caused it to suffer much more than the Whistling Swan *Cygnus columbianus*, which moved in larger flocks over open water.

The Mute Swan

Mute Swans *Cygnus olor* were royal birds in England and northwest Europe. There was an interesting dichotomy in their exploitation in these regions for, whilst carefully managed for profit in England (see below), they were hunted on a major scale in the southern Baltic when moulting and flightless. These hunts took place in late July and early August, with the birds pursued by hunters in boats before being shot or clubbed (Brusewitz 1969). Dawnay (1972) noted that such hunts were a royal prerogative, and recorded them from at least 1557 until about 1750 in Denmark. The birds were driven along the coast by fishermen for several days beforehand, while farmers on horseback prevented their escape inland. Bags of 200–500 were taken. In the Swedish part of the moulting area such hunting persisted into the second half of the 19th century (Brusewitz 1969).

In England, although originally an indigenous wild bird (Yalden & Albarella 2009), the species had, by the mediaeval period at least, become semi-domesticated, and the property of many owners. Ownership was recorded by swan marks made on the bill (particularly the upper mandible), leg or foot, or wing. Such swan marks were valuable property, in which there was a fairly brisk trade. Unmarked swans were the prerogative of the Crown. Theoretically no subject could own swans except through a grant from the Crown, but this limitation was widely disregarded so that ownership came to be largely restricted by wealth. This principle was enshrined in the Act for Swans of 1482, which laid down a property qualification for their ownership (see Ticehurst 1957 for a full account).

Abundance and legislation

One result of this regime was that Mute Swans became extremely abundant in England. Ticehurst (1957) made a very rough estimate of the population of cygnets and breeding adults in the Fens in 1553, where there were 800 registered owners of swans, based on data he noted as incomplete but which gave a total of 7,500 birds (2,000 adults and 5,500 cygnets). Ticehurst stated that to these would have to be added the cygnets of the previous three years, and calculated that this would bring the total swan population of the Fens to something in the region of 24,000 birds. He also estimated that around 700 swans were present on the River Lea in 1587.

Foreign visitors remarked on the abundance of swans on the River Thames. Ticehurst quoted the secretary of the Venetian Ambassador, writing to his master in 1496, that 'it is a truly beautiful thing to behold one or two thousand tame swans upon the River Thames as I, and also your magnificence, have seen'. A number of other observers also commented that they had never seen a river so thickly covered in swans. The Thames and the Fens were the two largest and most important swan-bearing areas in England. But there were large numbers elsewhere. Ticehurst quoted John Taylor on a voyage up the River Avon to Salisbury as reporting the presence of at least 2,000 swans. It is interesting to compare these figures with recent estimates of the total population of Mute Swans in Britain, which was put at 35,000 birds in 2002 by Delaney & Scott (2002).

The value of swans resulted in much petty crime, particularly swan-stealing, which eventually required legislation to control it. A commission of 1463 appointed to enquire into irregular practices on the Thames gives some idea of what went on. It was charged to look into 'the capture of swans and cygnets by hooks, nets, lymestrings and other engines and the taking of swans' eggs on the River Thames and its tributaries from Cirencester to its mouth and arrest and imprison the offenders' (Ticehurst 1957).

Swans were greatly esteemed as table birds and customarily eaten at Christmas while, as Ticehurst remarked, no large banquet was complete without them, sometimes in prodigious numbers. Swan-keeping was often the source of considerable profit and estates increasingly diversified into maintaining swanneries in the late 14th and 15th centuries, following the Black Death (Stone 2006). It was the cygnets that were generally eaten and, when swan-farming was a serious business, the cygnets were caught and kept in swan-pits, to be fattened on malt, barley or oats (Ticehurst 1957, Dawnay 1972). Dawnay noted that swan-farming for feathers or skins, for food or for sale as ornamental birds continued in the Netherlands until 1939.

All swan species were hunted on the breeding grounds when flightless in moult. Elaborate swan drives in Russia used sailing boats and nets, and often captured as many as 300 Whooper Swans in a day when flightless (Dawnay 1972). Dementiev et al. (1967) noted that human persecution and activities leading to habitat loss had reduced Whooper Swan populations alarmingly in Europe. No hunting was carried out in the Soviet Union when they wrote, and netting was banned. Bent (1925) also noted that the Whooper no longer bred in Greenland as a result of over-exploitation.

Feathers and skins

Trade in swans' feathers was a royal privilege in many European countries. Hunters and fisherman had to take live birds to the courts annually, where they were plucked and released

(Hobusch 1980). These were presumably Mute Swans, but Bewick's Swans *Cygnus bewickii* remained a common quarry in the Soviet Union where abundant, being much in demand for 'the great thickness of very beautiful snow-white down, which, when properly dressed by a London furrier, makes boas and other articles of ladies' dress of unrivalled beauty' (in Dawnay 1972). Brusewitz (1969) also noted that swan hunting in north Germany was considered commercially valuable for the yield of skins, used for bedroom mats or fur coats; down, used for powder puffs; and quills for pens. The meat was less valuable, and was sun dried or smoked and sold as breast of goose. Interestingly, Ticehurst (1957) does not mention trade in swan skins in England, although the eating of swans must have provided such by-products.

The Trumpeter Swan in North America was brought to the verge of extinction by the trade in swans' skins. Between 1820 and 1880 the Hudson's Bay Company sold 108,000 swan skins, mostly Trumpeters, in London alone (Dawnay 1972). The trade began in the 1770s and continued to 1900, being finally banned in 1918 (Ripley 1965). In contrast to Dawnay's figure, Ripley recorded that the Hudson's Bay Company handled only 17,000 swan skins, again mostly Trumpeters, between 1853 and 1877, suggesting a steep decline in the species was already in progress by the mid-19th century.

Other swans were also widely hunted for their skins. An observer from one of the American museums found 'hundreds of thousands' of Black-necked Swan *Cygnus melanocoryphus* skins in an Argentine warehouse awaiting shipment for manufacture into powder puffs (Chapman 1943). The trade was still flourishing in 1899, when the price quoted was 25 cents (about £2.00 today) per pelt (Gibson 1920).

Geese

Arctic native peoples have long hunted flightless and moulting geese (and indeed swans) by rounding them up and driving them into nets. Many communities relied on such hunting for their winter stocks of meat. Although nets were mainly used, the Icelanders developed permanent pens for catching moulting Pink-footed Geese (Scott & Fisher 1953), which appear to be unique.

Dementiev *et al.* (1967) noted that by the 20th century, the netting and shooting of moulting geese and the collection of their eggs in spring had become the primary cause of serious population decline in several species in the Soviet Union, particularly the Greylag (which had also suffered significant loss of breeding habitat to cultivation), the Greater White-fronted Goose *Anser albifrons*, the tundra race of the Bean Goose *Anser fabalis*, and the Brent Goose *Branta bernicla*. The Brent in particular was of great economic importance to Arctic peoples for food, and was in danger of extermination if hunting had continued to be unregulated. The great decline in wintering flocks of this species around the North Sea in the early- to mid–20th century must have stemmed partly, if not entirely, from this. In view of the adaptability in the goose's choice of winter food that the species has shown in the latter half of the 20th century, the earlier disappearance of the eel-grass (*Zostera*) beds from English estuaries seems of less significance. The effect of gross over-exploitation of the eggs of the Svalbard population of Brent Geese is described on p. 89.

With the introduction of conservation measures, notably the creation of refuges, the control of exploitation on the breeding grounds, and the control or banning of shooting in

Decade	Greater White-fronted Goose[1]	Barnacle Goose (Greenland)[2]	Barnacle Goose (Svalbard)	Barnacle Goose (Russia)	Dark-bellied Brent Goose	Pale-bellied Brent Goose[4]
1950s	10k–50k	5,120	1,003		16,500[3]	
1960s	50k–100k	10,580	3,625	25,200	27,500	9,771
1970s	200k–300k	18,860	5,200	43,665	73,130	12,727
1990s	400k–600k	40,000	23,000	267,000	300,000	20,000

Table 5.1. *Average total winter counts of Greater White-fronted, Barnacle and Brent Geese during 1950–1999 by decade in Britain and northwest Europe. Data from Ogilvie 1978 and Madsen et al. 1999.* [1] *Baltic–North Sea population only;* [2] *Figures for the 1950s–1970s were for peak counts on Islay. Only five complete counts of this population were attempted in this period, but these show a similar trend;* [3] *1956–1957 only;* [4] *Irish population only.*

the winter quarters, many European wildfowl populations have recovered. Table 5.1 gives examples of the wintering populations of Greater White-fronted Goose, Barnacle Goose *Branta leucopsis* and Brent Goose in Britain and northwest Europe. The Bean Goose has probably shown a similar pattern of increase, for winter numbers have more than trebled in the Netherlands over the same period, although there are now no significant wintering populations in Britain.

In the past, geese were taken in clap-nets in both Siberia and the Netherlands. In Siberia, Macpherson (1897) noted that the nets used were some 40m long and 2m wide. The birds were caught on permanent sites on riverside flats, and attracted to the nets by stuffed decoys and calls. As geese always rise into the wind, the catching area also had a long, deep net set to windward, which took great numbers as they rose. In the Netherlands the nets were usually set in shallow water, and geese attracted to them by live decoys.

Seebohm (1901) recorded that large numbers of geese were also harvested by Siberian hunters for their feathers and down, which were particularly valued for the warmth they gave by all Arctic peoples. He noted that when these birds retired into the tundra to moult, the Samoyed hunters had grand hunts amongst them, and returned laden with feathers and down, which they sold at the annual trading fair at Pinega.

Ducks

All ducks have suffered from being eaten over the centuries, although their palatability tends to vary with what and where they are eating. The only species I have come across which was apparently regarded as inedible was the extinct Labrador Duck *Camptorhynchus labradorius* (Greenway 1967). But the food value of some species, such as the Common Eider *Somateria mollissima*, may always have been secondary to the value of their feathers and down. Tastes also change, and species regarded today as not worth eating were once well-liked; the Common Shelduck, for example, which Edward Topsell, writing in the 17th century, noted 'was esteemed for the table as much as any waterfowl' (Harrison & Hoeniger 1972).

Figure 5.1. *Driving flightless wildfowl into nets in the 16th century. From Payne-Gallwey (1886).*

Use of nets

Large numbers of moulting ducks were taken when flightless in the Fens from at least mediaeval times. Ray (1678) illustrated a type of trap with long side-panels that guided the birds down into three tunnel-nets, from which they could be extracted (Figure 5.1). All such nets were temporary structures, erected where a drive was planned. Drives involving large numbers of men in boats and on shore pushed the ducks into these traps, and bags of up to 4,000 birds in one drive were recorded (Gladstone 1930) Such large numbers led to attempts to regulate the practice in the 13th, 16th and early 18th centuries (see p. 45).

The technique of driving moulting ducks into netting traps set around water bodies was practiced over a wide geographic area, and over a long period of time. It occurred not only in Europe but in North America, and in New Zealand, where Phillips (1922–26) recorded that 'the natives carefully guard certain lakes and engage in great duck drives, in which trained dogs are used. In the season of 1867 it is said that 7,000 were caught on one lake in three days'.

From the Middle Ages, all around the North Sea, duck traps or tunnels evolved from the simple netting of moulting birds described above. These involved a large hoop net erected at the end of a short canal or lagoon, which was dug from the shore. A close fence was erected on either side of the lagoon leading to the mouth of the net, behind which the operator could work unseen and control his water dogs, which went into the water through a trap door. Tame ducks were used as decoys to lure in wild birds, which were then driven up the tunnel to a catching-net by the dogs (Brusewitz 1969). Such traps were more or less permanent structures, and were still in use in northern Germany into the 19th century. Large numbers of duck were taken in the autumn and winter; Brusewitz records one site in north Germany where 67,000 ducks were taken in 1784. Such duck traps were not constructed in Britain, apparently.

Wildfowl were also widely taken in flight-nets and clap-nets. Flight-nets came in two basic forms: long nets, which acted much like the modern mist-net and were usually set to catch birds in the fowlers' absence, and drop-nets, which were operated by the fowler.

Long-nets were set to catch birds in flight, particularly at night. Set and left, they were cleared the following morning. In Lancashire, Mitchell (1885) noted that the nets were set when there was no moon, across the banks last covered by the tide (Figure 2.2). They were made of very fine cotton or linen thread, with a mesh of three to five inches (7.5–12.5cm), and were from 13 to 15 mesh deep, hung loosely on poles up to 20m apart to give plenty of bag. There was no limit to their length, and up to 750m might be set by one fowler. Geese, ducks, Curlew, Whimbrel, and all manner of other shoreline birds were taken. Similar nets were used on the Solway, the Wash, along the Lincolnshire coast and elsewhere in Britain, and in many different cultures around the world. Macpherson (1897) describes their use from continental Europe, China, Japan, Australia (where they were apparently imported by European settlers) and India.

Drop-nets were apparently not used in Britain. But before the steady improvement of guns in the 19th century, drop-nets were the usual way of taking sea ducks in the Baltic. These nets were up to 70m long, and were suspended on rings that slid along a cord, held taut on poles 3m or more high and suspended just above the water. The cord holding the net was operated by the fowler. The nets were set across narrow channels between islands and skerries, which the ducks, particularly Long-tailed Ducks, flew along in the early morning between roosting sites at sea and inshore feeding areas. Just before a flock of ducks struck the net, the fowler dropped it and enmeshed the birds (Brusewitz 1969). Their use was abandoned as no longer worthwhile in the late 19th century, by which time numbers of these sea ducks had been severely reduced. Similar nets were used in Siberia and by native American and Australian peoples (Macpherson 1897, Phillips 1922–26).

Another form of net was set to catching diving ducks. These were known as douker-nets in Lancashire, and set on flat frames at low tide over the feeding grounds to catch and drown the birds at high tide as they dived for food. Scaup and scoters were sometimes caught in considerable numbers on the Lancashire coast in the 19th century by this method (Mitchell 1885). Similar nets were also used in north Germany and North America, but their use in the latter was abandoned as it was found that the resultant waterlogging spoilt the birds' flavour (Phillips 1922–26, Bent 1923).

Clap-nets (see p. 35) have been used to take wildfowl for thousands of years. For ducks they were often set in shallow water, and live decoys were used to attract wild birds. As Markham (1621) advised his readers, such decoys also included herons staked beside the catching area, attached by a line to the fowler in his hide or hut, so that he could make them flutter occasionally. This ploy is illustrated in ancient Egyptian art, and several species of heron were used (see p. 51).

Snares

Snares for wildfowl were usually set in runs or arrays, suspended from a bar or thick cord across channels, for example, where ducks would swim into a line of them, or set along specially prepared strips for Teal *Anas crecca* (and also Snipe *Gallinago gallinago*). In Lancashire, such arrangements of snares were known as pantles; Mitchell (1885) gave a detailed description of their use and arrangement (see p. 127). Rather similar series of snares were used by Dutch and Scandinavian fowlers to catch geese. Macpherson (1892) described the wiles used to snare ducks in certain areas of Lakeland; snares were attached to a wooden frame fastened to the bank of a slow moving stream. Bait was scattered below and the ducks were snared as they up-ended to feed.

Duck decoys

Duck decoys (as opposed to decoy ducks) are pools into which ducks are attracted and then taken in nets. The true pipe decoy (Figure 5.2) almost certainly evolved from the duck-traps or tunnels used from the Middle Ages all around the North Sea, which in turn developed from the technique described above of netting flightless ducks. Pipe decoys were probably developed in the Netherlands from the 16th century, though they may have appeared around the same time in Germany, for Macpherson (1897) recorded that George I of Hesse had three decoys in 1575, for which he obtained decoy ducks from Friesland. There were clearly considerable numbers of decoys working in the Netherlands in the early 17th century. Southwell (1904) published details of a visit made there by Sir William Brereton, an English enthusiast for decoys, in 1634, specifically to examine the workings of the Dutch decoys, of which he visited up to 27 in three localities.

Duck decoys were a major source of wildfowl for the markets from the 17th century to the early 19th. They were managed mainly as commercial enterprises. But some, particularly cage decoys (see below), were constructed as house decoys, built as part of a country estate and worked to supply the house with wildfowl.

In England the first decoys were constructed in the 1620s, at Waxham in Norfolk by Sir William Woodhouse, and at Dodleston in Cheshire by Sir William Brereton (Southwell 1904, Coward & Oldham 1900). Knowledge of the construction and use of decoys came to England from the Netherlands. Brereton employed a Dutch decoy-man (as did Charles II in St James Park); Southwell suggested that this was not an unusual practice.

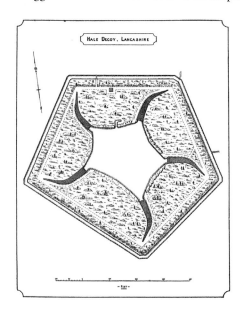

Figure 5.2. *Plan of Hale Decoy in Lancashire. This illustrates what is meant by a pipe decoy. From Mitchell (1885).*

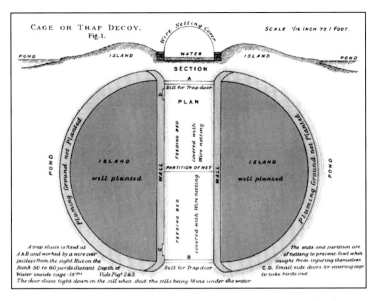

Figure 5.3. *Plan of the cage or trap decoy at Hardwick Hall in Derbyshire. This has been recently restored by the National Trust. From Payne-Gallwey (1886).*

Decoys were not confined to England, the Netherlands and Germany. They were also constructed in Wales, Ireland, Bohemia in the modern Czech Republic, France and Denmark (Kear 1990). Although it is usually stated that they were not constructed in America, Phillips (1922–26) found that pipe decoys of the Dutch type must have at least been attempted in 17th century Massachusetts. He quoted a reference to a general court held in Boston on September 6th 1638, which voted that one Emanuel Downing, who had imported the necessary materials

> *be given liberty to set up his duck-coy within the limits of Salem; all persons were forbidden to molest him in his experiments by shooting any gun within half a mile of the ponds where, by the regulations of the town, he should be allowed to place the decoys.*

The court afterwards granted other towns liberty to set up duck decoys, but they never prospered.

Early decoys in Britain often consisted of pipes placed around or at one end of large water bodies, such as the Broads in Norfolk, but they were never as successful as specially constructed pools. The most successful decoys consisted of a pond of one to two acres (*c.* one hectare or less) in extent, of variable shape, sheltered by woodland or other tall vegetation to limit disturbance and with three to eight (sometimes more) curved tunnel-nets, known as pipes, constructed along ditches radiating from the shore, each getting progressively narrower and lower until it terminated in a detachable funnel-net (Figure 5.2). Along the outer curve of each pipe, fences of reed panels were set in a chevron fashion, and joined by low jumps over which the decoy-man's dog leapt while at work (see below).

A second type of decoy was known as a cage or trap decoy, usually made as a house decoy to supply estate mansions with wildfowl. Figure 5.3 shows the plan of the cage decoy at Hardwick Hall in Derbyshire, which was of typical design. Ducks were enticed into the trap by feeding and by tame decoys, and the decoyman then lowered the trapdoors from his hut. Once lowered, the birds were held in the trap until the rest of the ducks on the pond had left to feed in the evening, when the trapped birds were bagged up. Heaton (2001) remarked that these cage traps were never as successful as pipe decoys. One at Haughton, Nottinghamshire, averaged only 300–400 ducks annually.

How decoys worked

There are many detailed accounts of how decoys were worked in the literature, for example Wentworth-Day (1954) and Kear (1990). Counter-intuitive though it may seem, a well-trained dog was generally used to attract ducks up the pipe. Most decoys also kept a flock of tame call-ducks, which the decoy-man summoned for food at the entrance of the pipe he wished to use by a whistle, attracting the wild birds on the pond to follow. Once they were in the entrance of the pipe the dog appeared, hopped over a jump and disappeared, to repeat the exercise further up the pipe. The wild ducks rushed after the dog in an attempt to mob it, and were drawn further in. Once the ducks were round the bend out of sight of the pond the decoy-man showed himself and drove the birds up into the funnel-net, in which they were finally caught. The great art of the decoyman was to do all this without disturbing the birds that remained on the pond. Lack of disturbance was considered essential, as ducks used these pools as daytime roosts.

The importance of such seclusion was recognised in law, although the degree of legal protection varied with the age of the decoy. After a period of 20 years without interruption by neighbouring landowners, the owner of a decoy enjoyed a right of action against anyone who wilfully disturbed or hindered his operations; case law was first established in 1706 (Marchington 1980). Regulations governing and preventing such disturbance were much more comprehensive in the Netherlands, where there were severe laws against trespass and disturbance of decoys, supported by long-established (and favourable) public opinion. In most cases this freedom from disturbance extended to a zone, of radius varying from 500–1,600m according to province, around the decoy and its wood; shooting was strictly forbidden in this zone (Matthews 1958).

(a)

(b)

Figure 5.4. *(a) Layout of reed panel fences along a decoy pipe, showing also the dog jumps and the decoy-man's dog at work. (b) The decoy man driving the ducks up the pipe once out of sight of the main pool. From Payne-Gallwey (1886).*

Decoys in Britain and Ireland

Payne-Gallwey (1886), in his history of decoys, listed 193 in England and Wales, and 22 in Ireland; more have been discovered since, and Appendix 2 (see p. 215) lists details of up to 237, nine of which were cage decoys, in England and Wales. None was finished in Scotland, although one was attempted. The actual dates of construction seem infrequently recorded. Sir Thomas Browne, writing of Norfolk about 1663, refers to 'the very many decoys, especially between Norwich and the sea' (Stevenson & Southwell 1890), suggesting that many of the Norfolk decoys were of this early date. But at least eight (26% of the known total) were built after 1810. In Lincolnshire and Essex few dates are known, but the records available suggest strongly that most of the 77 decoys in these two counties were in use in the first half of the 18th century, if not before. In Essex, Christy (1890) provided evidence that many were in operation before 1720. Altogether, of the 162 decoys for which I have found the details, 37 (23%) were in use or probably in use in the 17th century, 96 (59%) were in use in the 18th century up to 1810, after which drainage of major wetlands led many decoys to be abandoned, and 31 (19%) were constructed after 1810. This calculation assumes that those decoys marked * in Appendix 2 were of 17th-century date, and all but two of the 77 decoys in Essex and Lincolnshire were working in the 18th; three were certainly of 17th century origin.

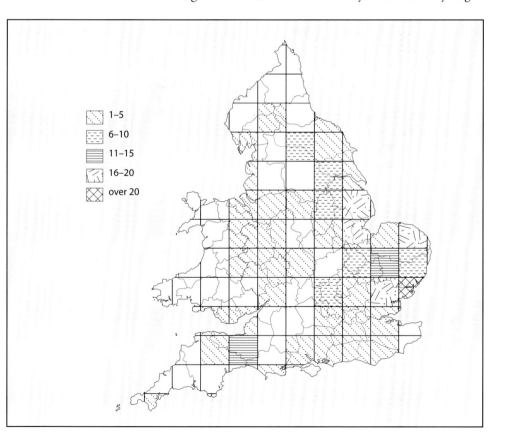

Figure 5.5. *Distribution of duck decoys in England and Wales by 50-km squares. County boundaries are pre-1974.*

Figure 5.5 illustrates the distribution of decoys in England and Wales by 50-km squares, showing the concentration in the major wetlands of eastern England, and a wide distribution in the lowland counties of England. A few small pools were constructed specifically to take Teal, a highly regarded table bird. One, perhaps two, of the Essex decoys were Teal ponds and there were others at Ivythorne on Sedgemoor in Somerset (Whitaker 1918), at Nacton and Orwell Park in Suffolk (Ticehurst 1932), at Kemsley in Kent (Ticehurst 1909) and at Peasmarsh in Sussex (Walpole-Bond 1938). Others no doubt existed.

Three of the Essex decoys started life as flight-ponds, designed for catching Pochard *Aythya ferina*, and then had decoy pipes added. Pochard and other diving ducks were regarded as a nuisance in decoys, as they dived out of the pipes and disrupted operations. Instead of pipes, Pochard pools had

> … *spring nets attached to long stout poles, weighted at one end so that by the removal of a peg they flew up and extended a net at a height of* [up to] *25 feet* [7.6 m]. *This was done when a gun was discharged. The Pochards, rising against the wind, the flight of poles in that direction being sprung, struck the net and fell down into the pens or pockets at the bottom, from which they were unable to rise, and were then secured (Glegg 1944).*

The pools usually had four sets of poles, to take advantage of different wind directions. Arthur Young, writing in 1807, recorded that a wagonload and two cartloads of Pochards were taken with a single rise of the net at Goldhanger decoy in Essex. The decoys at Brantham and Iken in Suffolk and Balby Carr, Doncaster, were also equipped with Pochard nets (Payne-Gallwey 1886, Ticehurst 1932, Limbert 1978).

Not knowing when many decoys were constructed means that it is difficult to say how many were working at any one time. But it seems fair to assume that their use peaked in the 18th and early 19th centuries, and that around 140 were then working in eastern England alone. Somerset was also an important county, with 14 or 15 decoys. Bearing in mind that up to 19% of known decoys were built after 1810, there were perhaps 180 or more in operation at the peak of the industry.

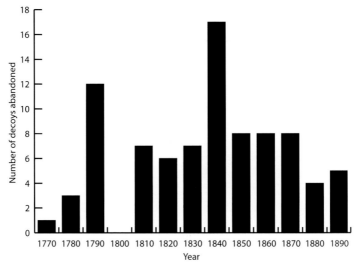

Figure 5.6. *The abandonment of English and Welsh decoys by decade, 1770–1900, as far as is known. Each decade is indicated by the first year. Note that the records are by no means complete.*

More is known about when decoys ceased operating. Figure 5.6 plots the number abandoned in England and Wales, for which dates are known, by decade from 1770 to 1900. Sites recorded as abandoned in the late 18th or early 19th century are recorded conventionally as 1790 or 1810. In addition, only three of 23 Irish decoys were still working by the 1880s, and most were probably abandoned before 1850 (Payne-Gallwey 1882). The main reasons decoys were abandoned were enclosure and drainage, and disturbance. Most of the Lincolnshire decoys were abandoned with the enclosure and drainage of the major wetlands there in the early 19th century. Wentworth-Day (1954), for example, recorded the loss of seven at Wainfleet between 1809 and 1829. Drainage was followed by major habitat change, from wetlands to arable farmland, with a steep decline in wildfowl numbers, particularly Mallard. Decoys swiftly became unprofitable (see Kear 1990). The import trade discussed on p. 87 contributed significantly to this decline in profitability, as it undercut local prices. Matthews (1958) gave agricultural improvement as the main reason for the steady decline in the numbers of Dutch decoys in the 19th and 20th centuries. Disturbance was particularly related to shooting, as wildfowling and the management of duck decoys were incompatible. This was the main reason that many of the Essex decoys, nearly all of which were sited around the major estuaries, were abandoned by the 1860s. Proprietors also came to prefer the sport of wildfowling at a time when decoys were becoming increasingly unprofitable. In the Netherlands, by contrast, large numbers of decoys continued in commercial operation into the 20th century, with 113 still operating in 1958, although this was half the total (220) active in 1838 (Mathews 1958).

Species taken in decoys

The principal species taken in English decoys were Mallard, Teal and Wigeon *Anas penelope*, which accounted for 37%, 32% and 29% respectively of the total bags for which there are sufficiently detailed lists (these are mainly for the 19th and 20th centuries, but worthwhile lists cover only 10 sites). Geographic position also influenced what was taken, with Wigeon being caught mainly on estuarine sites, particularly in Essex and Suffolk, where large numbers fed on the eel-grass beds in the estuaries, and used the decoys as secure daytime roosts; Mallard and Teal predominated in freshwater marshes. Species such as Pintail *Anas acuta* never figured prominently in English decoys.

The limited figures available do not suggest any marked change in the proportions of these species over the course of the 19th and 20th centuries. It is also instructive to compare these proportions with those recorded in the Netherlands for the 20th century by Matthews (1958). He found that the composition of the average bag for the whole country was Mallard 73%, Teal 15%, Wigeon 8% and Pintail 4%. The much higher proportion of Mallard perhaps helps to explain why decoys continued to be viable in the Netherlands long after they had been largely abandoned in England, for Mallard was the most valuable species.

It is impossible to calculate any convincing long-term trend for the average annual totals taken in English decoys. Information is scant and, as recorded by many authors starting with Pennant (1776), it was customary to record bags from decoys in dozens, which was how they were sold, counting Mallard as one duck but Wigeon and Teal as 'half ducks', confusingly counted as 18 to the 'dozen'. However, figures for the second half of the 19th and early 20th centuries suggest average annual bags of around 2,500 birds per decoy in England and Wales, and suggest little change over the period, a pattern supported by the detailed analysis of catches at Borough Fen decoy near Peterborough, dating from the late 18th century (Cook 1960). Such data that are available for the 18th century up to the 1830s suggest that a sharp decline in the average bag per decoy occurred during the 19th century. This is hardly surprising in view of the great habitat changes associated with enclosing and draining major wetlands, particularly between 1780 and 1850 (Shrubb 2003). This decline in average bags may have been by as much as 50%. Assuming that this calculation is valid, and that my suggested maximum of 180+ decoys working in the 18th century is correct, decoys were supplying nearly a million ducks to market in the 18th century.

Wildfowling

Wildfowling, the pursuit of ducks and geese with a gun for sport, was mainly developed in the 18th century, following the earlier development of the flintlock sporting gun, the first gun to make shooting flying birds practicable (Marchington 1980). Both crossbows and longbows had been used before this time and considerable skill might be demonstrated. It is reported that the Holy Roman Emperor Maximilian I once killed 100 ducks rising from the water from 104 shots with the crossbow (Brander 1971). Early users of the gun stalked their birds and shot them sitting, a procedure dictated by the unwieldy nature of early matchlock and wheel-lock guns.

In Britain, Peter Hawker did much to popularise wildfowling as a sport for the amateur in the early 19th century, particularly the art of punt-gunning. This was becoming established by the end of the 18th century, but most punt-gunners were professionals who made part of their living from shooting and supplying wildfowl to the markets, for which they were an important source. Coastal estuaries, and, prior to enclosure, inland fens and marshlands, all supported such professionals. Although Marchington (1980) considered that punt-guns evolved from the heavy 'bank guns' once used in the Fens, their ancestry perhaps also owed something to the 'barrow guns' used in Europe in pursuit of geese from the early 17th century. These were large-calibre shotguns mounted on wheeled carriages, which were trundled towards the target behind camouflage screens; Brusewitz (1969) illustrated one mounted through boards painted with a life-size grazing cow.

Punt-guns were formidable pieces of artillery. Payne-Gallwey (1882) gave a table of proportions, which showed total weights ranging from 60–180 pounds (27–81kg), bores from one and one-eighth to two inches (2.9–5.0cm), shot charges from 10–40 ounces (0.3–1.2kg) and barrel-lengths from seven to nine and a half feet (2.1–2.9 m), a list which was not necessarily exhaustive; some were double-barrelled. Most punt-gunners owned more than one gun, their size and capacity varying according to the conditions expected.

The successful management of a gunning-punt demanded a high level of skill, and it could be a dangerous sport in severe weather. The size of bags has tended to exaggeration, coloured by the large shots which were undoubtedly made and perhaps dominate the written record. Marchington (1980) recorded the bags made by Snowden Sleights, a professional fowler operating on the Derwent Ings in Yorkshire. Between 1891 and 1907 Sleights fired 830 shots with his punt-gun to obtain 5,352 head, an average of 52 shots a season for between six and seven birds per shot. Hawker, operating at Keyhaven and Poole Harbour in Hampshire between 1813 and 1850, bagged a total of 4,488 head of wildfowl, 78% of them Brent

Geese and Wigeon. In the 20th century William Mudge shot 15,228 Coot, ducks, Brent Geese and waders from a gunning-punt on Southampton Water and the Beaulieu River in Hampshire in 42 seasons, an average bag of 362 birds per season (Tubbs 1992). Bigger bags would probably have been made on less disturbed sites on the east coast of England and in Ireland in any comparable period. Thus Wentworth-Day (1949) described fleets of punts operating together for Brent Geese on the Essex coast. For example, in the winter of 1860 on the Blackwater, 32 punts under the leadership of Colonel Champion Russell killed 704 Brent Geese with one simultaneous discharge; about 250 more were picked up in the next few days. Several other very large shots were made around the same time using similar tactics. But these were exceptional events.

Overshooting

Both Mudge and Hawker commented on the large numbers of punt-gunners operating in the Hampshire Solent, and that they probably impeded each others' operations significantly. Complaints about overshooting in many British estuaries, particularly those near centres of human population, are commonplace in the 19th- and early 20th-century wildfowling literature. Perusal of 19th-century avifaunas leaves little doubt that such over-exploitation, combined with extensive habitat change, came to significantly affect wildfowl populations in Britain in the second half of the century. But over-exploitation was a general European and, particularly, an American problem. Several constant themes are evident in discussions in the American literature of the decline of wildfowl in this period. As in Europe, foremost among these was the prevalence of spring shooting, particularly of migrant breeding populations on their northward passage. This extended shooting in some States into May, and thus the breeding season, when it was most damaging to breeding populations. Grinnell (1901) also stressed the scale of indiscriminate shooting for the market in the United States, and

Figure 5.7. *The French system of duck-shooting by huttiers. From an original drawing by Peter Hawker (Diaries, December 1st, 1819).*

the improvement of guns; repeating shotguns became available in America in the 1890s, and automatics from 1904 (Doughty 1975). No check was imposed in America on large-scale commercial-market shooting by close seasons, game laws, game preserves or similar devices until late in the 19th century, when market shooting was progressively banned, close seasons enforced, quotas introduced for quarry species and bird protection legislation passed (see Chapter 3). Curtailing the shooting season by reintroducing proper close seasons and banning spring shooting were also important steps taken to correct over-exploitation in Britain and Europe.

Gunning-punts were little used in continental Europe, but an important source of supply there were the huttiers, wildfowlers who decoyed ducks to areas of open water commanded by a hide (or *hutte*) with tethered live call-ducks. The call-ducks were tethered in lines at right angles to the hides, and the wild birds shot on the lanes of open water between (Marchington 1980; see Figure 5.7).

Wildfowl in commerce

There was a considerable trade in wildfowl in Britain, such that the Bird Protection Act of 1880 described wildfowl as a staple article of food and commerce. As a broad generalisation, the main sources for markets were decoys in the 18th and early 19th centuries, and punt-gunners from around 1850. Christy (1890) recorded two tonnes of Brent Geese being sent from Maldon to London on one occasion by Colonel Champion Russell. Birds taken in decoys commanded premium prices. Most important provincial towns had dealers supplying local as well as metropolitan markets, and Bull (1888) and Macpherson (1892) recorded wildfowl being hawked around market towns by higglers (itinerant poultry-dealers). London was probably the biggest market. A statistical survey in the 1790 edition of the *Encyclopedia Britannica* calculated that London alone consumed '700 dozen of wild-fowl of several sorts' weekly for six months, a total of 200,000 birds (Kear 1990). Kear estimated that this could have absorbed the entire take of 100 decoys, although a proportion would certainly have been obtained from other sources, and the term 'wildfowl' may have included species other than ducks, particularly waders (Gladstone 1930). Nevertheless Defoe in the 1720s noted that

> it is incredible what quantities of wildfowl of all sorts, duck, mallard, teal, widgeon, etc. they take in those duckoys every week, during the season; it may be indeed guess'd at a little by this, that there is a duckoy not far from Ely, which pays to the landlord, Sir Tho. Hare, 500 l. [pounds] a year rent, besides the charge of maintaining of servants for the management; and from which duckoy alone they assured me at St. Ives, (a town on the Ouse, where the fowl they took was always brought to be sent to London;) that they generally sent up 3000 couple a week (Furbank et al. 1991).

Figures given by Mayhew (1860–62), although incomplete, suggested that the scale of this trade had declined significantly by the mid-19th century. He also noted that imports from Dutch decoys were then much more important than domestic sources.

One of the main problems of organising a trade in perishable goods such as birds was getting them to market reasonably fresh. Limitations in transport imposed by poor roads and uncertain sailing ships long restricted the reach and scale of such trade. As noted earlier, in the mediaeval and early modern periods this problem was met by keeping birds alive in stews, and transporting them to their destinations alive in cages or baskets, together with their food and water. Large numbers of Pintail were still being sent live from Dutch decoys to London in March and April in the late 19th century (Lilford 1895).

By the early 18th century, transport in England had improved sufficiently so that East Anglian decoys could serve the London markets. But it was the development of railways and steamships and, in the late 19th and early 20th centuries, of refrigerated transport, which allowed commerce in wildfowl to become international and global. Thus in Britain there was a considerable import trade in the 19th century, particularly from Dutch and Danish decoys and, in April and May, barrels of frozen birds from Siberia (Raven 1929). The construction of transcontinental railways facilitated the large internal trade in North America. Phillips (1922–26) estimated that no fewer than between six and ten million ducks, and possibly more, were harvested annually in the United States in the late 19th century, basing this figure on returns from markets, shooting clubs and State administrations. In the early 20th century 800,000 to 1,000,000 ducks a year were still being taken in California alone. A high proportion of these totals were taken by professionals shooting for the market, and large numbers were sent to city markets on the eastern seaboard. There was also a significant trade in birds sent to Europe in cold storage; Stubbs (1913) recorded that 'in past years large quantities of American gamebirds and wildfowl were sold fresh in the Manchester Market … but it is now a long time since I saw an American complexion on a game-dealers stock-in-trade, and perhaps this Transatlantic traffic has ceased'. This comment must reflect the restriction and eventual abolition of spring and market shooting in America, and the imposition of bag limits in the early 20th century.

European markets also received extensive imports from China, from the lower Yangtze (Phillips 1922–26), and there were major wildfowling industries in Egypt (continuing an ancient tradition), and in northern Iran on the southern shores of the Caspian Sea (Savage 1963). But how far there was an international element to this trade is unclear.

Considerable quantities of sea ducks were taken in the Baltic and along the north German coast, in all seasons. Brusewitz (1969) noted that they were an essential part of the coastal economy in the 19th century. The principal species taken were Greater Scaup *Aythya marila*, Common Eider, Velvet Scoter *Melanitta fusca*, Goldeneye *Bucephala clangula*, Red-breasted Merganser *Mergus serrator*, Goosander *Mergus merganser* and, above all, Long-tailed Duck *Clangula hyemalis*, which occurred in phenomenal numbers. Large numbers of birds, particularly Long-tailed Ducks, were shipped to the markets of Hamburg, Berlin, Leipzig and other inland cities (Phillips 1922–26). Until their numbers began to decline sharply in the late 19th century, the ducks were taken in flight-nets (see p. 77), and shot from hides among the skerries and islands and, on a massive scale, on the first patches of open water in the ice in the spring, where the ducks alighted in dense flocks and showed a fatal disregard of gunfire. On their breeding grounds in Scandinavia these species were also hunted when flightless with dogs and fishing spears (Brusewitz 1969).

Dementiev *et al.* (1967) gave considerable details on the economic importance of various wildfowl in the Soviet Union in the first half of the 20th century. Shooting for the market was an important commercial activity, and their accounts indicate that, in the Soviet era, this activity was at least partly regulated by State purchasing programmes, particularly in the steppe areas of western Siberia. But earlier accounts, for example by Seebohm (1901), confirm that market hunting was a long-standing tradition.

Feathers and down

Feathers and down were important wildfowl products. The trade in swan feathers was a royal privilege in many European countries (see above). The trade in the feathers of domestic geese is discussed in Chapter 1 (p. 26), and that in those of wild geese is discussed on p. 75. There is little reason to doubt that the feathers and down of wildfowl taken for the table was also preserved and used within the household. I can remember my mother doing this, with geese she reared and with ducks we shot.

The trade in sea ducks from the Baltic (discussed above) was important for the birds' feathers and down as well as for their meat. But the most important duck of all for this trade was the Common Eider *Somateria mollissima*. Eider down was and is highly prized for bedding, quilts and lightweight cold-weather clothing. Bent (1925) noted that, in the Nearctic, Inuit peoples made beautiful blankets from Eider skins for sale in Danish markets. They were made of the breasts of Eiders, from which all the feathers had been plucked, leaving the down on the skins. These were then cured, so they were very soft and pliable; the edges of these blankets were then trimmed with the cured skins of the heads of Common and King Eiders *Somateria spectabilis*. Bent remarked that they were the softest, lightest and warmest blankets he had ever encountered, and they were enormously expensive.

The following account of the eider down industry is summarised from Phillips (1922–26), Dementiev *et al.* (1967), Doughty (1979) and Hansen (2002). Although Iceland has the most highly developed eider down industry, down was traded in Norway, Denmark, Arctic Russia, the Faeroe Islands, Svalbard, Greenland and Canada, although only in the first three was the species systematically farmed as in Iceland. Attempts to set up an eider-down industry in northern Canada in the early 20th century were unsuccessful. Down was only collected from the Common Eider; the breeding populations of the other eider species and, indeed, of the Pacific race of the Common Eider are too dispersed to support systematic collection. Because of the value of down, Eider were nominally protected in all Danish possessions from at least 1702, although protection was not generally effective until the mid-19th century.

These regulations were never taken seriously in Greenland. There, wanton over-exploitation had wrecked Eider stocks by the mid-19th century. Large colonies have never re-established themselves since, as unremittingly intensive hunting and unregulated down- and egg-collection have prevented recovery (Hansen 2002). The decline in numbers over the 19th century can be shown by the sales of Eider down, which amounted to an annual average of 9,167 lbs (4,125kg) during 1822–1831. Calculations by Cott (1953) suggested that this represented the produce of around 110,000 nests. By the beginning of the 20th century sales amounted to an annual average of only 940 lbs (423kg), equivalent to 11,000 nests, a 90% decline. The species was similarly hunted for meat, eggs and down in Labrador and northeastern United States. A marked decline took place in much of the species' Nearctic range as a result.

Pennant (1776) noted that, in the 18th century, Eider down that originated in Hudson's Bay, Greenland, Iceland and Norway was imported into Britain through Denmark. But the trade was certainly older than that, for English merchants were trading in eiderdown from Iceland in the 15th and 16th centuries, and the Dutch were trading down from Arctic Russia in the 16th and 17th centuries. Eider colonies exploited for down in Russia have long been protected, and up to 452kg of down was collected annually from Novaya Zemlya alone in the 1930s, an important factor in the economy there. Nevertheless, outside such protected sites, Common Eider populations in Russia have been severely affected by shooting and egg-collection and have declined, although officially protected since 1931.

In Iceland, collection for domestic use probably dates back to the colonisation of the island in the 9th century, and early steps towards Eider conservation there were promulgated in the 13th century. In the past, Icelandic Eiders were exploited for their eggs as well as down, the custom being to take the first two clutches to eat or preserve when harvesting the down, with the third clutch being left to be incubated. Eggs are no longer taken today, and two down collections are made, one when incubation starts and the other when it has finished. Down production peaked in Iceland in the early 20th century, when a minimum total of 2,500kg was collected annually from around 250 colonies between 1909 and 1917. Production has declined since and, by the 1970s, it was about half the level of the early 20th century, from 200 colonies, many of them small.

Although the Eider was a favourite gamebird in 19th-century Norway, this country also had a down industry on some of the islands in its northern provinces. The lease-holder of the principal Eider island of Store Tamsoya was paying 48kg of clean down annually to the Government in the early 20th century. Protection was afforded to the species in the summer (and it is fully protected today), but down collection has now largely ceased. Norwegian hunters also collected large quantities of down from Svalbard in the late 19th century but no care was taken to protect the island's colonies; collection ceased by the second decade of the 20th century. Eider were afforded full protection on Svalbard in 1963, but no down is exported.

Eggs

The eggs of wildfowl have always been a valuable source of food, which perhaps, as Serjeantson (1988) suggested for seabirds, provided a substitute for domestic poultry in northern regions. Thirsk (2007) observed that the domestic rearing of wild birds, including ducks, by hatching eggs taken in spring to augment the family's meat supply, was a skill long practiced by ordinary folk.

All around the Arctic, wildfowl were exploited for their eggs, very often without any regard for the future of the populations concerned. The Brent Goose on Svalbard provides a clear example. The species bred there mainly on low, outlying islands, and was very vulnerable to eggers. Raids to take eggs by the crews of sealing sloops systematically swept up all the eggs, which, combined with indiscriminate shooting in the breeding season, drove the birds almost entirely away from many sites (F. C. R. Jourdain in Bent 1925). Such raids continued into the late 1930s, with hunters from Norway in particular coming to gather eggs for sale in Longyearbyen, Hammerfest and Tromso. They 'made their way from one suitable island to the next, systematically clearing each in turn of all eggs, without discriminating between Brents' and Eiders' eggs, nor even between those which are fresh and those which are incubated' (Cott 1953). Ogilvie (1978) noted that this population was so small and apparently declining as to raise concern for its continued existence.

Durnford (1874) noted that Shelduck *Tadorna tadorna* eggs were esteemed on the Friesian Islands, and the species had for many decades been encouraged to breed there in a semi-domesticated state. Where it abounded, as on the Island of Sylt, the natives made artificial burrows in the sand hillocks, cutting a hole over the nest chamber and roofing it with a sod to facilitate examination of the contents. Some hillocks contained 12–15 nests in a space of up to eight metres. The birds became tame enough to be taken by hand when sitting, and eggs were systematically gathered, leaving one or two to encourage continued laying, up to June 18th, at which point the ducks were left alone to rear a clutch.

Another species that was farmed for its eggs in Scandinavia was the Goldeneye. Its habit of nesting in hollow trees led to the custom of erecting nest boxes for it, to which it readily adapted. This facilitated the managed collection of the eggs. Brusewitz (1969) noted that this was an ancient practice in Sweden – the boxes were described by Linneaus in 1732 (Cott 1953) – and on 18th-century survey maps, Goldeneye boxes were sometimes used as boundary markers, emphasising their position as permanent features of the rural economy. Up to 20 eggs could be taken from a single female before she was left to rear a clutch.

Other wildfowl in the Baltic were less fortunate. Brusewitz (1969) noted that eggs were collected on a large scale on the Baltic islands, saying that 'as recently as the 1860s, the islands were so thoroughly stripped of all kinds of eggs of an eatable size each spring that the waterfowl, at any rate on the Swedish west coast, decreased in a most alarming manner. An eye witness summed up laconically: 'the water fowl are being wiped out to improve the taste of pancakes'.

But the duck most severely exploited in the north, once again, was the Common Eider. In the 19th century its eggs were taken throughout its breeding range, often on an unsustainable scale. Uncontrolled egging combined with breeding-season shooting came close to exterminating the species in the Nearctic, before protective measures were applied (Cott 1953, Doughty 1979). Eiders' eggs were an important dietary item in Iceland, but there they were carefully managed. In Svalbard the situation was similar to the chaos of the Nearctic; as with the Brent Goose, the main culprits were the Norwegians, who collected enormous numbers of eggs and down for sale in Longyearbyen and mainland Norway (Jourdain 1922, Cott 1953). In Norwegian Finnmark, Russian eggers were bringing quantities of Eiders' eggs for sale to Vardo and Vadso (Cott 1953), and eggs were still in great demand in Troms up to the late 1930s, resulting in a halving of the Eider population that had existed at the beginning of the 20th century (Cott 1953).

The examples I have discussed above include just some of the wildfowl that proved to be important egg-suppliers. Cott's (1953) account showed that more casual exploitation of this food source was a world-wide phenomenon.

CHAPTER 6
Gamebirds

Gamebirds have long been exploited, for the sport their pursuit provides and for their food value. The latter was the more significant before the 18th century, but the sport of shooting, with the great social occasions it involved in the 19th century and its commercial value in the 20th, came to be the more important thereafter. Nevertheless, there was always a commercial demand for game for the table, and attempts under the Game Laws in England in the 17th and 18th centuries to ban its sale to the public to preserve the landed gentry's privilege merely resulted in poaching serving this market, rather than legitimate sources.

Legally, gamebirds in Britain are defined as Red Grouse *Lagopus lagopus scotica*, Ptarmigan *Lagopus muta*, Black Grouse *Tetrao tetrix*, partridges and Common Pheasant *Phasianus colchicus*. Great Bustard *Otis tarda* was included in the legislation up to and including 1831, but would now be covered by the Protection of Birds Acts. The purpose of the English Game Laws as far as they affected birds from the mediaeval period to the 19th century was not their protection or conservation, but preserving the exclusive privilege of the landed gentry to hunt them. This was changed by the Game Reform Act of 1831, under which gamebirds became the legal property of the person upon whose land they were found (see Chapter 3 for a full discussion). That Act also laid down that hunting and trading in game required a game licence. The protection and conservation of gamebirds required separate legislation, laying down open and close seasons. These were first established in 1762, and the present seasons were largely established by 1773, and confirmed by the 1831 Act. As previously noted (see p. 44), Game Laws in Scotland differed in providing close seasons for gamebirds from the early 15th century. Legislation in the two countries was brought into line in 1772 and 1832.

Although taxonomically a gamebird, and occurring regularly in Britain as a migrant, the Quail *Coturnix coturnix* occupied an anomalous position in the Game Reform Act of 1831.

Whilst not legally defined as a gamebird, the Act nevertheless made trespass in pursuit of Quail an offence, and laid down that a game licence was required to take or kill them. The species was, however, included as game in the Irish Game Laws and in the Isle of Man, with a close season from January 1st to September 1st (12th in Man) (Marchant & Watkins 1897). Another migrant, the Corncrake *Crex crex*, was also taken for the table from at least the 16th century, and occupied the same anomalous position as the Quail under these Game Laws.

Trade

As with wildfowl, there was a significant commercial trade in gamebirds, legitimised after the Game Reform Act of 1831 but nevertheless widespread before. Mayhew (1861–62) gave some details of the numbers of gamebirds going through two main London markets in the early to mid-19th century. He recorded an average yearly supply of 57,000 grouse, 145,000 partridges and 64,000 Pheasants. To this he added some 39,000 birds in total sold by street hawkers of game. His account showed that game was available out of season, and that hawkers before 1831 were an important conduit for the activities of poachers, through the mail coach drivers (see p. 42).

Large quantities of gamebirds were also imported into English markets in the later 19th century. Gurney (1883) remarked that he was surprised that no mention was made in *The Zoologist* of the 'remarkable abundance … far exceeding the usual supply' of Black Grouse in every poulterer's stall in March of that year. Messrs Hunter & Son of Leadenhall Market told Gurney that they had sold up to 500 in February and nearly 350 in March, at six to seven shillings per brace, and that this was less than 10% of those sold through the market or sent to provincial markets throughout Britain. These birds were imported from Norway, Sweden, Russia and Germany, presumably on ice. Hazel Grouse *Bonasa bonasia* were also being imported from the same sources, and Gurney expected this trade to increase, reflecting their excellence as table birds. They sold readily at three to four shillings per brace. Gurney also recorded that at least one consignment of Grey Partridges *Perdix perdix* had been received from Russia.

Large numbers of Willow Grouse *Lagopus lagopus*, erroneously called Willow Ptarmigan by Yarrell (1843) and other early 19th century writers until corrected by Gurney (1883), were also being imported into London, mainly from Norway. Wilson (1840) recorded that 60,000 were snared in one parish in a single winter in Lapland, and Yarrell noted that one dealer would trade up to 50,000 in a season. He noted that in 1839 one dealer shipped 6,000 to London, 2,000 to Hull and 2,000 to Liverpool. These Norwegian gamebirds were sent frozen in the boats bringing lobsters into Britain. Wilson (1840) noted that large numbers of Capercaillie *Tetrao urogallus* were also traded from Norway into British markets.

Stubbs (1913) also drew attention to large quantities of American gamebirds being sold fresh in Manchester market in the late 19th century, the result of large scale market shooting in the United States. This trade died out in the early 20th century, presumably as the result of conservation measures imposed at source.

The following sections deal with each of the principal game species in Britain, plus Quail and Corncrake, both of considerable interest in this context.

The Red Grouse

The Red Grouse *Lagopus lagopus scotica* is an upland bird in Britain, its distribution largely coinciding with that of upland heather moorlands in the north Midlands, northern England, Wales and Scotland. In the past it was also found on lowland moors or mosses in Cheshire, Lancashire, Wales and East Yorkshire (Mitchell 1885, Coward & Oldham 1900, Nelson 1907, Lovegrove *et al.* 1994, Brown & Grice 2005). It has disappeared from most of these sites as a result of enclosure and conversion to arable land in the 19th century. Unsuccessful attempts were made in the 19th and early 20th centuries to establish populations on lowland heathlands in Suffolk, Norfolk, Surrey and Staffordshire (Babbington 1884–86, McAldowie 1893, Bucknill 1900, Harting 1901). The bird was introduced successfully onto Dartmoor and Exmoor in 1915–16 (Brown & Grice 2005), although a previous attempt to establish them there in 1820–25 had failed (Pidsley 1891).

The archaeological record shows a similar northerly and westerly distribution (Yalden & Albarella 2009). As with Black Grouse (see below) there have been Iron Age finds outside this range, at Meare in Somerset and Danebury Hill fort in Hampshire, and Roman finds at Ipswich in Suffolk and Great Staughton in Cambridgeshire. It seems likely that such records resulted from internal trade.

Red Grouse were sufficiently highly regarded as game in Scotland for close seasons to be set for them from the early 15th century, although these were subject to regular revision (see p. 44). Such seasons were not set in England until the 1760s (see p. 44). Thus Marmaduke Tunstall, an 18th-century ornithologist from Yorkshire, noted that, before this date, shooting started in June in north Yorkshire (in Nelson 1907) and MacPherson (1892) noted that the Naworth household accounts in the 17th century record that Red Grouse were regularly purchased in the breeding season.

These birds were almost certainly taken in snares or springes, set in the runs the grouse used when feeding, a method dating back to the mediaeval period (MacPherson 1897). MacPherson also noted that snaring grouse continued into the 19th century, and quoted the case of a Cumbrian moor, where the lessees employed a man to snare grouse on a large scale to pay the rent of the moor, which they then shot over whenever they had a spare afternoon. History does not reveal how long the grouse withstood this arrangement.

Grouse shooting as a sport seems not to have been much pursued before the second half of the 18th century. This seems implicit in Marmaduke Tunstall's observation, in 1784, that one reason for the decline of grouse in his day in north Yorkshire was 'the great improvement in late years in the art of shooting flying [birds]'. It is of interest here that Mackenzie (1949) noted that, in the past, lairds in northwest Scotland employed hunters to supply their establishments with deer and game, rather than troubling themselves with field sports.

Improvements in transport, particularly the development of a comprehensive rail network, opened up many previously remote areas to the whole battery of organised game shooting and game preservation. This was especially important for grouse-shooting in the uplands, particularly in Scotland. Vesey Fitzgerald (1946) noted that the development of shooting as a sport in Scotland was strongly influenced by the spread of the rail networks and greatly improved access; it attracted high rents. Sheep farming, already in economic difficulty, declined as shooting became valuable. Peter Hawker only went north to shoot grouse once, in 1812, and his diary records that the journey by coach from Andover to Glasgow took four or five days, although he extended this with stops *en route* for shooting and sight-seeing; the return journey from Glasgow to London took four days. Fifty years later the journey could have been accomplished inside a day; a century earlier and it would have taken four times as long (Blanning 2007).

Early shooters walked up their grouse, shooting the birds over pointers (see Plate 4). There were experiments in driving birds with beaters over a line of guns from around 1805, but this did not become general until the 1870s (Gladstone 1930).

By the end of the 19th century, grouse-shooting had become a considerable industry, and enormous bags were being recorded in England and Scotland. The record bag for Britain was 2,929 shot at Abbeystead, Lancashire, on August 12th 1915 (Gladstone 1930), and Gladstone listed four other bags of between two and three thousand birds, and at least 19 of between one and two thousand, virtually all of which were obtained in the first 15 years of the 20th century. Such bags are now a thing of the past (Tapper 1992).

The Black Grouse

Widespread in northern Europe and hill and montane areas of Central Europe, the Black Grouse *Tetrao tetrix* is confined in Britain today to marginal habitats in the uplands, where it has been declining, at least in some areas, since the mid-19th century. But at the beginning of that century it was found in suitable habitats virtually throughout Britain, except for a group of English counties in the east, and in the East Midlands (Figure 6.1).

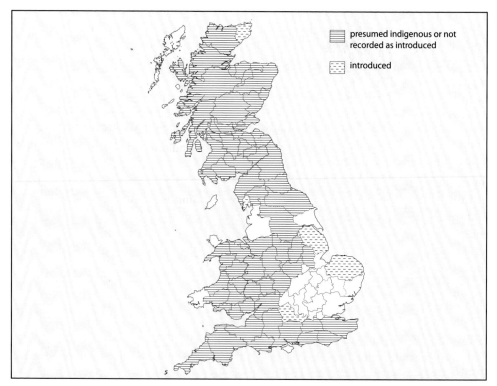

Figure 6.1. *The distribution of Black Grouse in Britain by pre-1974 county in the late 18th/early 19th century. Revised and redrawn from Shrubb (2003). Introduced Caithness in 1810 (Gladstone 1924). See also p. 95 and Table 6.2.*

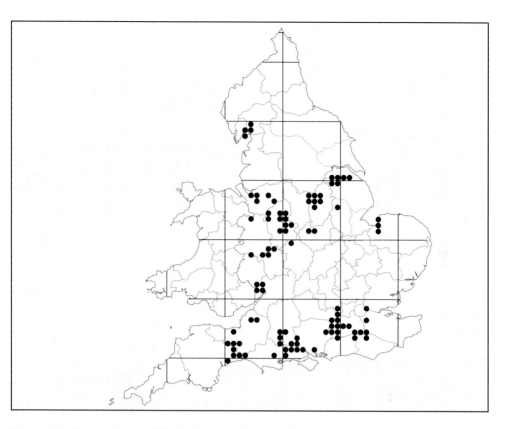

Figure 6.2. *The distribution of Black Grouse in the lowland zone of England in the early 19th century by 10-km square. Source 19th and early 20th century county avifaunas. The squares plotted are those into which sites given in the avifaunas fall.*

The 19th and early 20th-century county avifaunas indicate that the species was still common or abundant over most of its range until the mid-19th century; Millais (1909), for example, recorded autumn flocks of 200–300 birds in Scotland. However, populations in many Welsh counties were sustained only by releasing birds at various points in the 19th century. Forrest (1907) considered that only Montgomeryshire of the six north Wales counties he covered held a truly indigenous population. It is doubtful whether Black Grouse were present in Pembrokeshire after prehistoric times (Mathew 1894). But it was present on the Isle of Wight until the late 16th century (Cohen 1963).

Figure 6.2 illustrates the distribution of Black Grouse in the lowland zone of England, as defined by Rackham (1986), by 10-km squares, the squares plotted being those holding sites in which the species was recorded (other than as a vagrant) in the 19th and early 20th-century county avifaunas. This species disappeared from lowland Lancashire by the 19th century (Mitchell 1985). Table 6.1 summarises the more informative records available from these lowland areas.

Many of the more important sites plotted in Figure 6.2 were mediaeval chases or royal forests, such as the New Forest, the Forest of Dean, Cannock Chase, Sherwood Forest and the forests on the Hampshire/Surrey borders. 'Forests' in this context were areas where the king's, or

Devon	Clearly quite numerous on the commons of southeast Devon in first half of 19th century but only count was of 17 between Knapp and Gittisham in 1885 (Loyd 1929)
Dorset	Hawker and one other gun shot 11 brace in a day at Uddens in southeast Dorset in 1824. Tapper remarked that one of the best leks on the Hampshire/Dorset heaths was on the site where Bournemouth Town Hall now stands (Hawker Diaries, Tapper 1992).
Hampshire	Widespread on the lowland heaths and in 'great abundance' in the New Forest, although declining in 1791. 41 shot there in 1845–46 and 33 by two licenced guns in 1876–79. As there were 40–50 licencees shooting then, many more were probably shot but no records were kept (Tubbs 1968).
Gloucestershire	Four or five leks of up to 30–40 birds in the Forest of Dean in the 1850s (Swaine 1982).
Nottinghamshire	Found all over the heathy areas of Sherwood Forest and plentiful in Kersal Forest and at Coleorton Corner. Whitaker's father shot 16 in a day in Mansfield Forest in the 1860s, and packs of 50 were seen up to the 1870s (Sterland 1869, Whitaker 1909).
Staffordshire	Abundant on Cannock Chase until the end of the 19th century. The record English bag of 252 shot in a day was made here in about 1860, and several other bags of more than 100 were made in the same period. Bags of 30–40 were still being made at the end of the century (Gladstone 1930, Smith 1930–38).
Surrey	A large stock on the Surrey heaths until the late 19th century. Up to 20 seen on Leith Hill in 1832. A rapid decline after 1870 (Borrer 1891, Bucknill 1900).
Sussex	Present all along the forest ridges in the north of the county, and very common all over Ashdown Forest until mid-19th century (Turner 1862, Borrer 1891).

Table 6.1. *Numbers of Black Grouse in lowland England before the mid-19th century.*

in some cases great magnate's, deer were preserved. As such they were subject to Forest Law, the full application of which inhibited agricultural expansion and improvement (Cantor 1982a). They were areas of mixed habitats, but particularly of heathland for the deer, and had multiple uses. They were also extensive, with an average area of 5,000 acres (2,024 ha) (Rackham 1986).

As Black Grouse are birds of the interface between moorland or heathland and forest (Tucker & Heath 1994), the creation of the forest system was possibly beneficial to them; the presence of forestal rights, Forest Law and forest courts from the 11th century gave added protection to extensive continuous areas of suitable habitat, which were already afforded some protection from development due to their low fertility. However, it is not at all clear that lowland England was occupied on any scale during the mediaeval period. For example, Yapp (1983), in his survey of gamebirds in mediaeval England based on illustrations in mediaeval manuscripts and definitions in early vocabularies, concluded that Black Grouse, although present, were not well known. Except for household accounts for establishments in the upland zone of England and in Scotland, the species was also absent from the lists upon which Table 1.3 (see p. 20) is based until the mid-17th century, when it occurs in the household accounts of the Dacres at Herstmonceux Castle (Lennard 1905). Similarly, Black Grouse are absent from the price lists published for the London market up to the end of the 16th century by Jones (1965). But they were recorded as being taken by falconers in Wiltshire in the 17th century (Brentnall 1947).

Bedfordshire	A total of 143 continental birds released at Woburn between 1901 and 1907; a few pairs bred in each year without establishing a permanent population (Fitter 1959).
Berkshire	Introduced into Windsor Great Park in the 1860s, from where they colonised Ascot Heath, Woking Common, Bagshot Heath and other suitable sites (Clark Kennedy 1868).
Cheshire	Attempts to re-establish populations at Newchurch Common and Abbot's Moss in 1901 failed (Guest *et al.* 1992).
Hampshire	Released into Woolmer Forest in 1780s and again in 19th century to augment existing but declining populations. Also releases in the New Forest in 19th and early 20th centuries for the same purpose (Kelsall & Munn 1905, Tubbs 1968, Tapper 1992). Released at Hurstbourne in late 19th century (Fitter 1959).
Lancashire	Introduced into Furness between 1832 and 1840 (MacPherson 1892).
Lincolnshire	Probably introduced to the heaths of the northwest (Smith & Cornwallis 1955). Harting (1901) mentions releases at Frodingham and Caistor in 1871–72.
Norfolk	Probably introduced into northwest Norfolk between King's Lynn and Hunstanton, but details unknown. Persisted there until late 19th century. Attempts to establish birds on the Beeston Hills near Cromer in the mid-19th century and at Two Mile Bottom near Thetford between 1885 and 1887 and again between 1897 and 1901 were unsuccessful, although a few pairs were recorded as breeding on Thetford Warren in 1907 (Stevenson 1866, Riviere 1930).
Suffolk	Introduced at Elvedon in 1865 but soon died out. Thereafter repeated and unsuccessful attempts to establish the species at this locality. 20 released on heathland east of Ipswich in 1899 also failed to establish themselves (Ticehurst 1932).
Surrey	Releases on Leith Hill in early 19th century and at Hurtmore in 1815 to augment existing but declining populations (Bucknill 1900).
Sussex	Introduced or reintroduced at Hollycombe near Linchmere in mid-19th century, and onto Black Down in 1840. Died out in the latter area by 1890 (Borrer 1891).

Table 6.2. *Introductions and releases of Black Grouse into lowland England in the 19th and early 20th centuries. From county avifaunas and Fitter (1959).*

Black Grouse were only listed from sites in Cumbria, Northumberland, Durham and Yorkshire in Parker's (1988) analysis of archaeological records for 89 sites in Roman Britain. Yalden & Albarella (2009) covered more Roman and Iron Age sites, but, for a species they noted as well-represented in the archaeological record, they found only two southern records for the Iron Age. These were at Meare in Somerset and Danebury hill fort in Hampshire, neither that far from heathland occupied by Black Grouse in the 19th century. In the Roman era, Yalden & Albarella noted no record south of Staffordshire.

These points raise the possibility that many of the known lowland English populations may have established themselves. It is otherwise difficult to believe that a species as highly regarded for the table as the Black Grouse was in later centuries should have been ignored, particularly as it was certainly exploited in the mediaeval and early modern periods in northern England and Scotland. In Scotland, Black Grouse were protected with a close season from as early as 1427 (see p. 44); Hector Boece, in 1526, described them as 'not unlike … a pheasant, both in quantity and savour of their flesh'. Their market price was fixed at 6d in 1551 (Baxter & Rintoul 1953).

There was a considerable trade in Black Grouse in London in the 19th century, when Yarrell (1845) recorded that the supply of the birds to the London markets was 'very large and continuous from the end of August until the following month of April; during the first four months from Scotland, and afterwards from Norway and Sweden'. Although by the fourth edition of Yarrell (Yarrell & Saunders 1882–84) this trade was no longer mentioned, Gurney (1883) showed that it still continued and had probably increased.

Reintroductions

Whilst there may well have been early introductions or releases on southern English heathlands of which we now have no record, a feature of the 19th century accounts for Black Grouse is the frequency with which it was shot in lowland England, in places well outside its established range, indicating a marked tendency to wander. Such a tendency would provide a mechanism for the species to spread into new areas from existing centres. There were, however, numerous introductions or releases into lowland heathland in England in the 19th century, largely to obtain or maintain stock for shooting; these are summarised in Table 6.2. Their history suggests that some areas of heathland in England were less suitable than was then believed. In particular, persistent attempts to establish populations in Norfolk and Suffolk were uniformly unsuccessful outside northwest Norfolk, perhaps because the areas concerned were too dry.

Lowland English populations of Black Grouse declined rapidly in the second half of the 19th century, with the enclosure of the commons and heaths (see Shrubb 2003). The need of Black Grouse for high levels of habitat diversity made the fragmentation of large areas of heathland, which resulted from the enclosure of different parishes or commons at different times (see Tubbs 1985), particularly significant. Diversity declines with fragmentation, greatly increasing the chance of any remaining fragment losing essential habitat elements. Such factors affected the species throughout its lowland range in northwest Europe, where more than 90% of lowland heaths in Denmark, Sweden and the Netherlands has been lost since the mid-18th century (Webb 1986).

Upland populations have also declined, largely as a result of habitat deterioration caused by agricultural intensification, afforestation and increased stocking rates of sheep (and deer) (Baines & Hudson 1995). Marmaduke Tunstall, writing in 1784, ascribed the decline in his day to similar causes, and to the great increase of skill in shooting flying birds (Nelson 1907). Many of the 19th century avifaunas also stressed the importance of overshooting in the decline of the Black Grouse, particularly shooting too early in the season, when broods of young were especially vulnerable, and from the activities of poachers. Poachers often shot them when roosting (Coward & Oldham 1900), and Nelson (1907) quoted Graves's *British Ornithology* (1813) that poachers took considerable numbers of males by imitating the call of the hens, presumably in spring, with up to 50 males being taken by this means in two days. MacPherson (1897) described another ploy, where poachers concealed themselves in a hide constructed of oat sheaves, provided with a perch for the birds to alight on to feed. Being fond of feeding on oat stooks, the grouse flew in and could, with care, be seized by hand.

Black Grouse exploitation in Scandinavia

In Scandinavia, the Black Grouse was the most economically important gamebird. It was heavily exploited, with large numbers sold through the markets, particularly in the 18th century. For example, in Sweden, 3,038 were shipped from Sundsvall to Stockholm in 1749, and in 1760 around 80,000 were sold in Stockholm and 200,000 sent to other cities (Brusewitz 1969). Brusewitz noted that in the mid-18th century people became sick of eating such gamebirds, and whole cartloads of Hazel Grouse and Black Grouse remained unsold in the markets. They were shot over pointers and, in good Black Grouse country, two or three guns had no real difficulty in shooting 100 birds per day in the early 19th century. The grouse were also shot from hides at their leks, and unrestricted shooting with improved guns at such sites in the early 19th century caused serious population declines. In winter, the birds were also shot over painted wooden decoys, mounted in the birch trees the grouse were feeding in. This method became very popular around many provincial Swedish towns and, since one gun could readily kill 20 in a morning, numbers again suffered severely (Brusewitz 1969).

In more northerly areas, from Scandinavia into Siberia, trapping was the more usual way of taking Black Grouse, either by snares set in bushes and baited with twigs of birch buds, or in 'Blackcock baskets', tall cone-shaped baskets with a pivoting perch across the top, which tipped the bird into the trap when it tried to reach the bait. The bottom of the basket was set with spruce branches which prevented its escape (MacPherson 1897). Broods of young were also taken in nets. Another common way of netting Black Grouse was to take them while roosting with a pole-net and a lantern. In winter, Black Grouse roost by plunging into the snow from a tree, and then burrowing deeper. The warmth of the bird's body melts the snow. Roosting birds were betrayed by depressions in the snow above them. After locating a roost-site, the fowler pressed his pole-net down hard on the sleeping grouse, and caught it. Brusewitz (1969) noted that this method of taking Black Grouse was recorded back to the mediaeval period.

The Capercaillie

The archaeological record shows that the Capercaillie *Tetrao urogallus* was widely distributed in Britain in the past, occurring as far south as Somerset in the Roman era (Yalden & Albarella 2009). Its later history in England is obfuscated by the confusion over names to which Yapp (1983) drew attention (see p. 110). Yapp's account, however, indicates that Capercaillies were present in English woods into at least the 11th century; this is supported by archaeological records from Anglo-Saxon Durham and York (Yalden & Albarella 2009). There were also 12th- and 13th-century records for Leicester and York, and a post-mediaeval record for York in the archaeological record, and Yapp (1983) concluded from the evidence of illustrations in mediaeval manuscripts that Capercaillies were present in English woods until around 1300. However, some later archaeological records may have resulted from importations. There was certainly long-distance trade in Capercaillies in Europe in the 17th century; Ray (1678) recorded many being sent to the markets of Padua and Venice from the Alps in 1665.

Many early authors attest to the excellence of the Capercaillie as a table bird. Ray (1678), for example, described its flesh 'as of a delicate taste and wholesome nourishment', an opinion shared by Chapman (1897). Ray also remarked that 'it seems to be born only for Princes and great mens tables'. It certainly figured in important feasts in Scotland in the 16th century. They were apparently also readily domesticated. Yarrell (1845) noted that this was often done in Sweden, the birds being kept, sometimes for long periods, in aviaries built for the purpose: 'These birds were so perfectly tame as to feed out of the hand. Their food principally consisted of oats and of the leaves of the Scotch fir ... they were also supplied with [an] abundance of native berries when procurable'.

Capercaillie persisted in Scotland until the late 18th century, the last birds being shot in 1785 on Deeside (Pennie 1950–51), but they were becoming rare by the early 17th century, when it was forbidden to buy or sell them (see p. 45). The species also disappeared from Ireland in the mid-18th century, although it had been afforded legal protection in an Act of the Irish Parliament for the Better Preservation of Game in 1711 (Hall 1982). The main reason for the extinction of this species from England, Ireland and Scotland was the loss of suitable woodland habitat, largely to agriculture but, in Scotland, also to climatic deterioration (Rackham 1983, Smout 2000). However, not all ancient pine woodland was lost in Scotland, so exploitation may have contributed; this is difficult to assess, as there seems to be little published information on the hunting of this bird in Britain in the past, although Hall (1982) quoted Thomas Fuller in the 17th century, referring to Ireland, that it was easy to shoot.

Catching Capercaillies

A common way of taking Capercaillies practiced right across Scandinavia, northern Russia and into Siberia was the use of baited deadfall traps. Birds were also taken in snares set in their runs and, if required alive, in bag- or purse-nets similarly set. As both deadfalls and snares were used for Red Grouse over most of its British range (MacPherson 1897), these methods could well have been used there for Capercaillie.

Capercaillies were ruthlessly extirpated in northern Europe with the development of improved sporting guns from the beginning of the 19th century; they were extinct in many of their haunts in southern Sweden by 1820. Two features of this activity were particularly damaging: the excessive shooting of displaying males in spring, and of hens and their half grown broods, decoyed to the gun by calls, in August. They were also shot at roosts at night on a fairly large scale in both Scandinavia and central Europe, and blasted out of the tree-tops when feeding in winter. In good years such feeding flocks could comprise 150–200 birds, nearly all cocks, and it was not unusual for a single gun to kill the whole flock. An ancient custom, still practiced in northern Europe, was to shoot the birds over dogs known as tree-barkers. These were trained to find and flush Capercaillie (and Black Grouse), follow them, and stand barking near the tree in which the bird alighted, thus holding it so that the hunter could come up and shoot it (Brusewitz 1969).

Reintroduction

With the extensive planting of new pine woods in Scotland from the second half of the 18th century, suitable habitat for Capercaillies was re-established, and the bird was reintroduced into Perthshire from Scandinavian stock from 1837. From there it spread into woodlands over much of eastern and central Scotland, reaching Sutherland by the end of the century. They were strictly protected and, where shot, were usually driven like pheasants. Gladstone (1930) noted bags of up to 60 in a day, with a record of 69 shot at Dunkeld in 1910.

Pennie (1950–51) made the interesting point that Capercaillies were afforded no protection or close season under the Game Laws in force at the time of writing, which had been promulgated after the species became extinct in the 18th century. They were regarded as pests by the Forestry Commission because of perceived damage to young plantations, the young shoots of pine being an important food for the birds. As a result they were shot by foresters in all seasons, and nests were destroyed. Capercaillies were finally afforded a close season from February 1st to September 30th by the Wildlife and Countryside Act 1981, and a voluntary ban on shooting was introduced in 1990.

The Grey Partridge

Although replaced by *Alectoris* species in parts of the south, the Grey Partridge *Perdix perdix* was the most important gamebird of lowland farmlands in Europe until well into the 19th century. They have always been highly regarded for their sporting qualities and excellence as a table bird (Potts 1986), and commanded high prices in the mediaeval period, averaging 4d (well above a labourer's daily wage) in London markets in the 14th century. Prices in London rose consistently in the 16th century, averaging 7^{1}/2d, and reaching one shilling by the early 17th (Jones 1965). Their sale was banned in 1603 in an attempt to bolster the Game Laws, but there is little evidence that this early ban was observed.

The archaeological record shows that Grey Partridges were common and widespread in Britain in the mediaeval and early modern periods, where it is one of the most frequently recorded species (Yalden & Albarella 2009). It is also one of the most frequently recorded and numerous in the sources used in compiling Table 1.3 (p. 20). The development and expansion of open-field arable systems from the late Saxon period to the mid-14th century, by which time they had come to occupy between a quarter and a third of the area of England (Rowley 1982, Rackham 1986), together with the expansion of cereal farming in the 12th and 13th centuries (eg Cantor 1982b), also makes it likely that Grey Partridge populations increased significantly in the mediaeval period. These changes to land use would have created much ideal new habitat for them. However, partridges were probably never as abundant as they became in the 19th and early 20th centuries.

Before the development of the flintlock fowling piece, which made shooting flying birds practicable, the main methods of taking partridges were with nets and with falcons or hawks. The birds were also shot with early guns and with crossbows, usually on the ground and with the aid of stalking horses. In the mediaeval period, some aristocratic and gentry establishments employed partridge-catchers; for example, at Framlingham Castle, Suffolk, in the late 13th century, the partridge-catcher kept falcons and eight dogs for this purpose (Stone 2006). Stone also recorded that the Reeve of Chevington paid 3/3d to two men for catching 50 partridges (and 12 pheasants) in 1347–48, the birds being sent to London for the Abbot of Bury St Edmund's.

Falconry and netting

Falcons were flown over setting dogs and spaniels. Setting dogs were trained to locate the game and lie down once it was found, causing it to freeze and stay put. The falcons were then launched to wait on high above the scene of action; the spaniels flushed the game, and the falcons stooped. The falcons most widely used were Peregrines *Falco peregrinus*; a good bird could make up to six flights in a day. Accipiters, most usually male Goshawks *Accipiter gentilis*, were also used, and were flown directly from the fist. Considerable numbers of partridges could be taken in a season. For example, Hobusch (1980) recorded that Margrave Wilhelm Friedrich of Bayreuth took 14,087 partridges with falcons in 25 years up to 1755.

The most common form of netting was the use of a drag-net; this also involved setting dogs to hold the birds, allowing the hunters to draw the net over them (Figure 6.3). If holding cover was poor, a hawk-shaped kite or a hawk itself was flown to hold the birds. Other forms of net commonly used included long-nets, similar to those used for larks (see p. 163), and tunnel-nets. The latter comprised netting-wings that guided the birds into a tunnel of netting, with a catching net at the end. The area in front of this arrangement was baited to encourage partridges (and also Quail) to feed there. The birds were then gently driven into the tunnel by the fowler, who was disguised in a cow's skin or using a stalking horse (Figure 6.4). Illustrations look improbable, but contemporary descriptions leave little room to doubt that this was done.

Partridges were also snared, and Haines (1907) described an interesting device used in Rutland. The hunter was armed with a wand around 15 feet (4.5m) long, with a noose of twisted horse hair at the end. Thus armed, the hunter stalked the partridge in the long stubble and, slipping the noose over its head, secured it.

Figure 6.3. *Catching partridges with a trammel or drag-net and a setting dog. From Ray (1678).*

Figure 6.4. *Catching partridges with a tunnel-net. From Ray (1678).*

Pictures and descriptions of these sports leave one with the distinct impression that partridges were tamer before the 18th century than they subsequently became or are today, perhaps because this kind of hunting would often have removed the entire covey. Arthur Berger (in Brusewitz 1969) suggested that this caused inherited changes in partridge behaviour. Those which squatted tightly were taken, leaving those that did not to pass on timidity and flight as a characteristic in the population.

Home-rearing

Another way of obtaining partridges was to take the eggs and rear the young in the poultry yard. Thirsk (2007) suggested that this was widely done, and not just in gentry and yeomen's households. Household accounts record food being bought for partridges (and Pheasants), sometimes in considerable quantities. Thus Woolgar (2006) recorded that the Earl of Oxford paid for wheat, barley and seven quarters of oats (1.07 tonnes) for this purpose in 1431–32. The poaching forays in Sussex parks in the 14th century described by Cooper (1863) (see p. 110) were probably raiding such yards. In the 17th century, James VI/I made considerable efforts to improve partridge stocks on the royal estates by releasing birds taken from other areas (Macgregor 1989). This was a common practice in game management in the 19th century and beyond.

Bag sizes

Partridges were certainly abundant on the plains of Europe in the 18th century, and considerable bags were made in France from the second half of the 17th century; Pepys recorded 300 being shot on the plain de Versaille by the King of France and his entourage in 1666 (diary, March 21st 1666). Much larger bags were made in the 18th century. For example the Ducs de Berry and de Bourgogne shot more than 1,500 partridges on the plain of Saint-Denis on June 30th 1706, many of which must have been little more than half-grown, and the king and his entourage shot 1,700 there on September 13th 1738 (Blanning 2007), whilst 2,593 (and 1,593 hares) were shot at Chantilly in two days in October 1785 (Gladstone 1930). Such bags were achieved by shooting with up to six guns and loaders, a performance which must have required considerable precision. While these bags may have been exceptional, bags of several hundred were not, which argues for a thriving and abundant partridge population.

Considerably smaller bags were the norm in Britain in this period. But one needs to be careful in using shooting bags as an index of population in gamebirds in the past, since the methods of, attitudes to and legal basis of shooting for sport all influenced this. Thus Gladstone (1930) pointed out that, while these very large bags of partridges were being made on the continent in the 18th century, such bags were then unfashionable in Britain. Underlying this was a significant difference in the legal basis of taking game. By the 18th century, the right to kill game had become far more widely diffused in British society than on mainland Europe; in Britain it rested on a property qualification, possession of which conferred the right to kill game anywhere, a factor which inhibited creating the conditions necessary for taking large bags. Hawker's diary suggests that this system had led to significant overshooting by the early 19th century. For example, on September 1st 1819, he recorded 'the coveys were wilder than ever I yet saw them in the first part of the season … I had four shooting parties round me, and the best half of my ground was beaten before I took the field'. And again on September 1st 1821 'The corn being so much in the way this season, I had made every attempt to postpone the shooting, but to no effect; and no sooner was it daylight than old Payne and his son, two vagabonds under the toleration of Mr. Widmore, were popping away before my house'. On the continent, particularly in France and Germany, shooting game was the exclusive privilege of royal and aristocratic elites, and this remained the case until the political upheavals of the mid-19th century.

From the early 19th century, partridge shooting passed through three distinct stages in Britain, from shooting over pointers (see Plate 4), to walking up in line, to driving (see p. 105). Shooting over pointers needed good holding cover, provided in the 18th and early 19th centuries by the long stubbles left by reaping with the sickle, and by root crops where they were still sown by hand. Potts (1986) remarked that such shooters often walked a mile for each bird shot, and Newton (1896) commented that shooting over pointers came to an end in little more than six weeks from the start of the season, on September 1st.

There are comparatively few records of partridge bags made from shooting over dogs, but Hawker kept detailed accounts of everything he shot. Shooting over dogs for 49 seasons from 1802 he averaged 143 partridges per season, his seasonal bags varying from 10 (in a season in which he did not shoot in September) to 388. His average daily bag in 25 Septembers was 24 birds. As the century progressed his bags tended to decline, and his diary is full of complaints about the difficulty of getting near enough to the birds to shoot successfully. Lilford (1895) recorded that 15–20 brace was regarded as a fair day's bag for two guns shooting over pointers with muzzle loaders.

As the 19th century progressed, enclosure, the development and spread of crop rotation (see Shrubb 2003), and the increasingly close preservation of game, particularly the destruction of predators (Tapper 1992), led to a significant increase in the population densities of partridges, whilst increasingly restricting the activities of traditional qualified sportsmen. At the same time improvements in sporting guns, particularly the invention of the percussion cap and then of early breech-loaders, greatly increased the rapidity and certainty of firing, raising the potential for making sizable bags. With these changes the method of shooting partridges by walking in line (walking up) emerged, where the birds were first walked into holding cover such as roots or thick stands of clover, and then shot by guns walking in line with the beaters.

Some very large bags were obtained by this method. Gladstone (1930) recorded that bags of 628 partridges shot in a day at Buckenham, Norfolk, in 1858, and 664 shot in a day there in 1859, were the records for England. But larger bags were obtained in Scotland, with a record of 753 shot by 10 guns at Panmure, Angus, on September 20th 1870, and 533 and 687 shot on the following two days.

Walking in line remained the principle method of shooting partridges in Britain until late in the 19th century and remains so in Europe today (Potts 1986). But experiments in driving partridges over the guns by beaters walking in line started in 1845 in East Anglia, although this did not become widespread until the 1870s or 1880s. Its successful adoption depended largely on the development of efficient breech-loading guns and their ammunition. It resulted in further significant increases in the size of bags obtained, and around the turn of the 20th century it was not unusual for bags of 1,000 in a day to be obtained on the best partridge manors of eastern and southern England (Gladstone 1930). Such tallies were eclipsed by those obtained on European estates, where some huge bags were obtained in the 19th century. But it was often the practice there to release birds netted on outlying beats of the estate into the area where shooting was to commence (Gladstone 1930). Bags of up to 3,000 birds in a day were not uncommon on large estates.

Although the record British bag of 2,015 partridges in a day was made at Rothwell, Lincolnshire in 1952, partridge bags were broadly in decline for much of the 20th century, particularly after 1945 (Tapper 1992). In the west, particularly Wales, the decline was marked by the early 1900s (Shrubb 2003), and was associated with the change from arable farming to pastoral and the decline of gamekeeping that followed. More generally, the partridge population in Britain collapsed in the 1950s and early 1960s as a result of the revolution in farming methods, particularly the widespread use of herbicides (Potts 2012). The bird has never fully recovered and is unlikely to do so, although Grey Partridges remain relatively widespread members of the farmland avifauna.

The Red-legged Partridge

It is unclear why Red-legged Partridges were ever introduced into England. Fitter (1959) suggested it was 'to reinforce the overshot stock of native Grey Partridges'. But this was hardly likely when the 19th century saw continued releases of birds into a greatly increased Grey Partridge population. I think the most likely reason for the introduction of *Alectoris rufa* was that it was something different, and quite decorative.

The Red-legged Partridge occurs naturally in Iberia, southern France, northwest Italy and Corsica today, though it formerly bred further north in France into southwest Germany, from whence it disappeared in the 16th century, and Switzerland, where it died out in the early 20th century. Long (1981) ascribed these populations to early introductions, but Voous (1960) considered them to be part of the species' original European range, from which it withdrew during the climatic deterioration of the Little Ice Age. The species was certainly introduced to Madeira before 1450, to the Azores in the 18th century and probably to the Canary Islands. Attempts to re-establish the bird in Switzerland and to introduce it to several central European countries have failed (Voous 1960, Bannerman 1963, Glutz von Blotzheim *et al.* 1973, Long 1981).

Introduction to Britain

In Britain, the first attempt to introduce Red-legged Partridges was made by Charles II, who released birds imported from France into the park at Windsor in 1673. Other early releases were made at Wimbledon between 1721 and 1729 – the birds were all destroyed by disgruntled neighbours; at St. Osyth in Essex in around 1768; in Harting, Sussex in 1776; and in Suffolk in the 1770s and 1790s (see Appendix 3). It is normally considered that the last mentioned were the basis of the present English population, but I doubt this, for there were frequent releases of fresh birds throughout the 19th century (Appendix 3) and contemporaries considered releases on the Norfolk/Suffolk borders in the 1820s as more significant (Pycraft 1936). The fact that so many releases were made over so wide an area, virtually all by shooting interests, suggests strongly that no single release provided the sole basis for a national population.

Having introduced Red-legged Partridges, many sporting estates found that they proved unsatisfactory for walking up from cover (see p. 105), being difficult to flush and running like hares. Gamekeepers also believed they drove out Grey Partridges, and spoiled their nests. Subsequently, determined but fruitless attempts were made on many sporting estates to exterminate them. Attitudes changed from the mid-1840s, when it was discovered that Red-legged Partridges provided excellent shooting as driven birds (p. 105), which led to the spate of introductions in the second half of the 19th century.

Today, this species is much the most common partridge in the English countryside; partridge shooting everywhere in Britain depends very largely on its presence, and often on continued releases of reared birds.

The Quail

Quail *Coturnix coturnix* are unusual among gamebirds in being long-distance migrants, breeding across Eurasia but wintering in Africa and India (Voous 1960). Although recorded quite widely in the archaeological record of Britain (Yalden & Albarella 2009) and in the lists upon which Table 1.3 (see p. 20) is based, Quail were possibly never very abundant in Britain. They were highly esteemed for the table and large numbers were consumed at major feasts, but many of these may have been imported. This certainly seems to be the case for the Neville feast of 1465 (Table 1.5, p. 22), where the total of 1,200 Quail provided were recorded as '100 dozen', the unit in which imported birds were marketed. Nevertheless, Thirsk (2007) recorded that Lord Leicester, when Chancellor of Oxford University, had a supply of Quails sent from Brill, in Bernwood Forest, Buckinghamshire, when giving a great feast in Oxford in September 1570. Thirsk also noted that there was something of a fashion for breeding

Quail among the gentry of the eastern counties around 1600, which may also point to some difficulty in obtaining wild birds, and Ray (1678) recorded them as being 'somewhat rare with us in England'. James Hart, a 17th-century physician, noted that, of game for the table, Quails were then the height of fashion, which he described as a fad for something rare (Thirsk 2007).

Viscount Lisle

The Quail's relative scarcity in England is also suggested by the Lisle letters (Byrne & Boland 1983, Bourne 1999). Viscount Lisle was Deputy Governor of Calais during 1533–1540. He organised what amounted to a considerable export trade in Quail to England, the birds being caught in very large numbers around Calais and in Flanders. As a great delicacy, they were sent as presents to members of the royal family and other important persons, and to people to whom Lisle or his wife owed money. As Byrne & Boland noted, such gifts were an essential part of the system of patronage, and were sent 'to sweeten requests, attract attention or to turn away wrath'. The birds were sent live in cages, together with food and water, and recipients tended to insist on their being fat. Consignments of up to 45 dozen were involved, and at one point a Calais poulterer was pressing Lisle for payment for 43 dozen.

The scale of Lisle's activities and the evident value placed upon these gifts suggest that the birds were not readily obtained in any numbers in England – it seems unlikely that such gifts would have been so highly regarded had this not been so. Another indication of the Quail's relative scarcity in England lies in the prices paid. In England, 16th-century documents show that the price of Quail varied from 2d to 4d each – in Calais Lisle was paying 2d per dozen. It is also possible that the scale of trapping carried out in northern France and Flanders contributed to the species' limited status in England. Prices had increased considerably in the 17th century when, for example, Lord Dacre, at Herstmonceux Castle, paid 24/- for 48 Quail purchased from a French ship in the 1640s (Lennard 1905).

Lisle's activities were not the only factor in the trade in Quail between England and France in this period. For the provisions for the meeting between Francis I and Henry VIII at the Field of the Cloth of Gold in 1520, over 3,000 Quail were supplied (Bourne 1999b), while another 2,784 were supplied for the meeting of these two kings and the Count of Flanders at Calais in 1532 (Bourne 1981). Poulterers were also importing Quail into London around this time, which Hope (1990) noted were imported from Flanders live in baskets with their food and water.

The Quail trade continued into the 19th century, when Yarrell (1845) recorded that the bird-catchers of France decoyed hundreds of dozens into their nets, using Quail calls to attract the birds to them, a traditional method of great antiquity, long practiced in France. Yarrell noted that the males arrived first, and the trappers imitated the call of the female to entice the birds into their nets. He remarked that on examination of dozens together in their cages it was rare to find a female. These birds were brought by French Quail-dealers to the London market, which consumed large numbers – up to 3,000 dozen in a London season (May to August), when they were in particular demand.

Hunting and status in the British Isles

The 19th- and early 20th-century county and national avifaunas give a fairly clear picture of the Quail's distribution and status in Britain around the turn of the 19th century. It was locally common or abundant in rough grassland areas, particularly at the edge of the Fens, in Cambridgeshire, Suffolk, Norfolk, and Lincolnshire, and in Hertfordshire and Lancashire, but otherwise it was an uncommon but regular and generally distributed summer visitor. In Scotland it was more common in the west than the east, and was locally numerous in some areas, for example Wigtonshire (Adair 1892). However, there are few records of their being hunted on any scale in Britain. Most of the avifaunas recorded a few being shot during September partridge-shooting. Colonel Hawker, for example, only shot a total of 58 in 51 seasons up to 1853 (diaries). Nevertheless, Booth (1881–87) noted that bevies (family parties of quail) reared on the fenlands between Newmarket and Cambridge gave fair sport early in the season, with up to seven brace being obtained by two guns in half a day.

The only reference I have found in the county avifaunas to Quail being netted is in Nelson (1907), who recorded that they used to be taken in nets at Bridlington, Yorkshire in the 1840s. Ray (1678) recorded that they were netted in the standing corn by fowlers attracting them to the nets by calling on a Quail pipe, and they were also taken in tunnel-nets like partridges (Figure 6.5). White (letter to Pennant, 1768) noted that 'Quail crowd to our southern coast and are often killed in numbers by people that go on purpose' but he did not say how. That practice, if ever common, had apparently ceased by the 19th century.

Contrary to the general pattern in Britain and northern Europe in the 19th century, Quail in Ireland showed a considerable increase from early in the century. The bird was scarcer in unreclaimed areas of the west but otherwise abounded in inland and maritime counties. Ussher & Warren (1900) wrote:

Figure 6.5. *Catching Quail with a tunnel-net. Note the caged call-birds as decoys. From Ray (1678).*

The extensive growth of wheat, leaving stubbles full of weeds, in the early part of the century, and the multiplication of potato gardens up to the time of the famine, were facts in favour of this bird, and bevies of Quails used to be met with commonly through the winter months in every cultivated district; this species was then considered more numerous than the Partridge, and appeared to be resident from its abundance at all seasons.

Many were shot, and significant numbers were sold in Belfast market in the winter, and exported to Glasgow in the early spring (Gray 1871). The population collapsed after the famine of 1846–48, as much of the arable land was converted to pasture or reverted to moor. By 1880, Quail were regarded as virtually extinct in Ireland, with only a few records annually (Ussher & Warren 1900).

The Mediterranean

All the 19th- and early 20th-century avifaunas record a sharp decline during the 19th century, which contemporaries ascribed to changes in agricultural habitats and methods, particularly enclosure and the loss of the rough grasslands of the fen edge. It is more likely, however, that this marked and universal decline was the result of a considerable increase in the number of Quail taken for the market around the Mediterranean, particularly during spring migration. This was an ancient tradition recorded, for example, in the Old Testament (Numbers 11: 31–32), and frequently illustrated in ancient Egyptian art. But there was a substantial increase in such exploitation in the 19th century, particularly in Egypt, where Nicholson (1951) noted that an export of 300,000 live Quail from Alexandria to Europe rose to two million in 1897 and more than three million in 1920. Most of these birds were consigned to France through Marseilles, although the final destination of many was England. The scale of this trade affected the numbers taken elsewhere in the Mediterranean. For example, on Capri, the numbers taken annually during spring migration fell from 150,000 in a good year in the 1850s to *c*.30,000 by 1904 (Nicholson 1951).

There were two main reasons for this massive increase in the Quail trade. First, with the development of steam ships, railways and ultimately refrigerated transport, the transport of perishable goods became more rapid and more certain, extending the reach of trade in commodities such as Quail. Most of this trade passed through Marseilles, which handled eight million Quail destined for France and England, the main consuming countries, in 1895 (Hobusch 1980). Secondly Goodman & Meininger (1989) noted that, until the late 19th century, netting in Egypt had largely been for local consumption; with British occupation from 1882, trapping was intensified and export to Europe was encouraged. Hundreds of thousands were also shot in Italy on spring migration and sent from ports such as Brindisi and Messina to Marseilles (Nicholson 1951). Lilford (1895) recorded seeing a steamer at Messina in May 1874 carrying 6,000 pairs of Quail alive in cages, all destined for the London market.

During the 1930s the numbers of Quail exported from Egypt annually declined to 285,000–580,000, more than half taken in spring; the same pattern of decline was repeated all along the north African coast. In 1937, taking Quail in the spring was banned in Egypt and French North Africa, and the import of live Quail into Britain and France was prohibited before July; the trade collapsed completely with the outbreak of the Second World War (Moreau 1951). Some recovery of the British population was evident as a result (Nicholson

1951). Today, considerable numbers are still taken in Egypt in both spring and autumn (Goodman & Meininger 1989), and doubtless elsewhere round the Mediterranean; it is doubtful whether the species will ever regain its former abundance, although marked 'Quail years' still occur (Gibbons & Dudley 1993).

The Pheasant

Introduced to most of western Europe, the Common Pheasant *Phasianus colchicus* has long been prized as a table bird. It was much favoured by the Romans, who probably carried it to Italy, France and Germany, rearing them like chickens for the table (Hill & Robertson 1988). Both Parker (1988) and Yalden & Albarella (2009) recorded Pheasants in Roman archaeological sites in Britain and, as elsewhere in western Europe, their early history there was entirely as a bird of the poultry yard.

Yapp (1983) pointed out that the word *fasianus* in early mediaeval Latin documents did not mean pheasant, as once assumed, but what we now call the Capercaillie, and he traced the first use of the word for what we call 'Pheasant' to as recently as the 15th century; prior to this, the bird was referred to as *fesaund*, or variants thereof. Yapp concluded from his study of early illustrated manuscripts, dictionaries and vocabularies that the Pheasant was neither well-known nor feral in England before the 14th century.

Rearing and establishment of Pheasants

Nevertheless, there are indications that Pheasants were being fairly widely kept by that time. For example, Cooper (1863) recorded a series of prosecutions in the mid-14th century in Sussex for poaching forays in parks, warrens and chases to take deer, rabbits, hares, partridges and pheasants. At least 14 sites were involved, and the context suggests that the gamebirds were being kept in pens or enclosures.

Pheasants occur regularly in the household accounts of gentry establishments in the later mediaeval and early modern periods (Table 1.3, p. 20), often with details of expenses for food for them, again indicating that they were being hand-reared. They were reared at Henry VIII's palace of Eltham in the 16th century (Simon 1952). Large numbers were sometimes provided for major feasts. Henry III requisitioned 290 for his Christmas feast in 1251 (see p. 23), 200 were obtained for the Neville feast in 1465 (Table 1.5, p. 22) and 888 were supplied for the meeting between Francis I of France, Henry VIII and the Count of Flanders at Calais in 1532 (Bourne 1981). Pheasants were always expensive to buy, the average market price in London ranging from 4d in the late 13th century to 2/8d in the late 16th (Jones 1965). Jones's price lists also indicate that the birds were available throughout the year, except during Lent, another indication that they were managed as poultry rather than game. No proper close season appears to have been applied until the mid-18th century, when an Act of 1762 made it illegal to take them (or even have them in one's possession) between February 1st and October 1st (Gladstone 1930), a close season that still applies.

The rearing of Pheasants as poultry continued into the 17th century; Thirsk (2007) noted, as an interesting aside on their management, that the Earl of Rutland bought his poultry-woman scissors to clip their wings in 1607. By this time, however, Pheasants were presumably also established in the open countryside as feral birds, at least in southern and eastern England. For example, the Le Strange accounts from Hunstanton recorded Pheasant being taken with hawks in the early 16th century, and expenses for maintaining Goshawks to take them. The Act of 1762 mentioned above also exempted birds taken in the wild in the open season and kept for breeding, suggesting that feral birds were being used to stock the poultry yards; there is little evidence that they were being reared for shooting at this time (Gladstone 1930).

Ray (1678) described Pheasants as woodland birds, preferring coppiced woodland and feeding on acorns, berries, grain and seeds. He noted that the common ways of taking them were with hawks or with nets and setting dogs, as for partridges. Nevertheless, Pheasants were not generally distributed as feral or wild birds in England before the late 18th century, when Pennant (1776) included them only in his section on domestic poultry, and were not a primary gamebird until the 19th century (Hill & Robertson 1988). Hawker (1893), for example, shot only 754 in 51 seasons from 1802, compared to more than 7,000 partridges, and his diary indicated that the appearance of a Pheasant at Longparish was sufficiently unusual to be the signal for everyone to turn out in pursuit.

Pheasants had presumably spread into Scotland by the late 16th century, when an Act of the Scottish Parliament of 1594 forbade anyone from killing them in or within one mile of the king's woods or parks. In the late 17th century, they were still being reared at some noblemen's houses (Baxter & Rintoul 1953). They remained scarce until the 19th century. Muirhead (1895), for example, found no county record for Berwickshire before 1810, although most landed proprietors were preserving them by the mid-19th century and there was large-scale rearing for shooting by the late 1880s. This seems to have been the pattern for much of Scotland, and in the north and west the population was probably always dependent on rearing. This also appeared to be case in parts of west and northwest England (see for example Bull 1888, MacPherson 1892, McAldowie 1893, Coward & Oldham 1900).

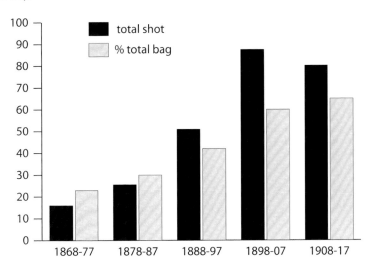

Figure 6.6. *The total number of Pheasants shot on five Welsh estates by ten-year periods from 1868, and their percentage of the total bag of game shot. The estates were Bodorgan (Anglesey), Voelas (Denbighshire), Stackpole (Pembrokeshire), Tythegston (Glamorgan) and Powys and Lymore (Montgomeryshire). For 1868–77 only the first three sites are included. Pheasants shot are in thousands. Data from Matheson 1963.*

Figure 6.7. *Game card from Thirkleby Park, designed by Sir Ralph Payne-Gallwey, recording the bag from November 1903.*

In Wales, the Pheasant was introduced to Pembrokeshire in the late 16th century, and spread thinly into other parts of that county thereafter. There was also an early introduction into Meirionnydd in the same period. No further information about these early introductions is available and, outside one or two large estates, Pheasants remained quite unfamiliar over most of Wales until the 19th century, when there was a great increase in rearing and preserving them from the second half. The effect this had on shooting bags is summarised in Figure 6.6.

This pattern was repeated on estates throughout Britain, with vast numbers of Pheasants being hand-reared and shot, particularly between about 1880 and 1914. Gladstone (1930) remarked on the reticence of landowners in publicising their bags, but noted at least 12 estates where bags of 2,000–3,000 in a day were obtained; the record English bag was 3,937 shot at Hall Barn, Beaconsfield, Buckinghamshire, in December 1913. Nevertheless, most estates gave shooting guests a game card, setting out the results of shoots; Figure 6.7 illustrates one from Thirkleby Park.

It is difficult to disagree with Rintoul & Baxter's (1935) judgement that the Pheasant could hardly be classed as a genuinely wild bird. It is still reared in large numbers, with more being reared and released than are shot (Tapper 1992). In effect, it has remained a bird of the poultry yard throughout its history in Britain.

The Corncrake

Whilst including this species among the gamebirds may seem unusual, the Corncrake (or Land Rail) *Crex crex* was highly esteemed as a table bird from at least the 16th century, when Dr Muffet wrote that 'Railes of the land deserve to be placed next to the Partridg, for their flesh is as good as their feeding good and they are not without cause preferred to Noblemens Tables' (Yarrell & Saunders 1882, Mullens 1912). This bird had the same anomalous status under the 1831 Game Reform Act as the Quail. Thus, whilst not legally defined as gamebirds, trespass in pursuit of Corncrakes was an offence under the Act, and a game licence was required to take or kill them. In Ireland and on the Isle of Man, close seasons were laid down, in Ireland from January 10th to September 1st and in Man from January 1st to October 1st.

Gladstone (1943) shows that Corncrakes were sold in the markets from at least the mid-18th century. But despite Dr Muffet's favourable opinion of them as table birds, they do not appear in the lists upon which Table 1.3 (see p. 20) is based, nor in the market lists given by Jones (1965). This might be the result of confusion over names, for Pennant (1776) in Appendix VII suggested that the word *Rey* or *Ree*, which does occur in household accounts and market lists, should be translated as the Corncrake (*i.e.* 'rail'), as opposed to Ruff or reeve; Pennant (1776) remarked that in the 15th century the latter were not much in vogue, a view supported by the paucity of records in mediaeval household accounts and market lists.

Considerable numbers of Corncrakes were shot during partridge-shoots in the 19th century, usually when walking up partridges in September. Bags of up to 15 brace in a day were quite widely recorded in the 19th century avifaunas, and larger bags were reported. For example, Ticehurst (1909) recorded 211 Corncrakes being shot at Acryse Park, Folkestone, during August and September 1880, and Gladstone (1930) listed three bags of 50 or more in a day, and reported that a Mr D. Darell of Hillfield House, Devon shot 263 in 1919; altogether, this gentleman shot 1,461 between 1909 and 1929, an average of 70 a year. Corncrakes were also shot on spring migration, mainly in the north, and on their breeding grounds, the birds being attracted to the gun with calls; they were shot while crossing specially mown tracks in the herbage (Bull 1888). Imitating the call of the Corncrake to draw the birds to their doom has history in Britain; Ray (1678) noted that fowlers in his day imitated the call by drawing a knife through dry wood. They then snared or trapped them in standing corn.

CHAPTER 7
Waders

Most of the common wader species have been exploited for food over the centuries, and some, such as Dotterel *Charadrius morinellus*, were also hunted for their feathers; Dotterel feathers were used for tying fishing flies. Wader eggs were also widely collected, particularly those of the Lapwing *Vanellus vanellus* and, in the Netherlands, the Black-tailed Godwit *Limosa limosa*. In some areas the eggs of other waders, mainly Ruff *Philomachus pugnax* and Common Redshank *Tringa totanus*, were gathered with and sold as Lapwing eggs (Lubbock 1845) as, later, were Black-headed Gull *Larus ridibundus* eggs. Virtually all the waders, but particularly Dotterel and other plovers, Ruff, Snipe *Gallinago gallinago*, Woodcock *Scolopax rusticola*, Black-tailed Godwit and Curlew *Numenius arquata* and, among shorebirds, the Knot *Calidris canutus,* were regarded as great delicacies on the table.

Early household accounts and market lists show that plovers, Snipe, Woodcock and Curlew were the species most frequently accounted for, particularly plovers and Woodcock. This pattern perhaps arose because these species were widely distributed inland in both summer and winter, making them more easily accessible than birds of the shore, rather than from any specific preference. It is often difficult to accurately assess the importance of individual species in these accounts, since the names or terms used may differ from those in common use today. Thus the term *stynts* covered small shorebirds generally, *Snyte* was used for the Snipe, but may also have been used for small shorebirds. Golden Plover *Pluvialis apricaria* was known as Grey or Green Plover, and the Lapwing was known as the *Wype* or *Lapwynke,* and sometimes as the Bastard Plover; a common name for it in the 19th century was Green Plover, a name which occasionally occurred earlier, for example in 1633 (Simon 1952).

As noted in Chapter 1, waders were often fattened up in mews or stews. The Ruff is perhaps the best known example of this, but it was often done with Black-tailed Godwit, and also practiced with Lapwing, Curlew, Knot, Dunlin *Calidris alpina* and perhaps Dotterel. Waders were also readily tamed. Lapwings, often originally injured, were often kept as pets in gardens to control slugs and other undesirables (Shrubb 2007). Brusewitz (1969) recorded that during his journey through Skåne in Sweden, Linnaeus saw flocks of tame Ruffs walking about with chickens and geese outside the byres at Marsvinsholm Castle. In Tudor England, Anne Boleyn kept Dotterel, sent as presents for her table, in her garden until required, as did Jane Seymour (Byrne & Boland 1983, Bourne 1999b).

Many waders were expensive, and prices rose sharply after the 16th century, when attempts to regulate those charged by poulterers were progressively abandoned, and factors of supply and demand left to set them (Jones 1965). Table 7.1 lists the prices charged for the most significant wader species from the 13th to the first half of the 20th century. Prices given in household accounts tend to be lower than those shown for markets, perhaps partly because some of the former were a *douceur* to the bearer of gifts, rather than the actual price for the bird. Gladstone (1943) also shows that Stone Curlew occasionally appeared in the markets, priced at four shillings, in the early 19th century. All the species tabulated were available in the markets into the early 20th century, except Dotterel.

Small shore birds were usually sold in dozens but only rarely appear in price lists, perhaps subsumed in the category 'great birds', although this term is usually taken to mean the larger passerines. The price lists in Jones (1965) also show that some waders, plovers and Curlew, for example, were offered throughout the year, confirming the lack of a close season.

As the 19th century progressed a further demand for waders came from collectors. Whilst mainly concerning rarities, species such as Knot or Spotted Redshank *Tringa erythropus* were much sought after when in summer plumage (see Plate 7), and inland occurrences of species normally associated with the coast were regularly shot and mounted as specimens for local collections.

Species	14th century	15th century	16th century	17th century	18th century	19th century	20th century
Plovers	2		2½	6	16	18–30	18–30
Eurasian Dotterel			2¼		18	36	
Red Knot			2½		36	36–48	2–3
Ruff			2	30	36	15–60	8–15
Common Snipe	1		1½	4	12	24–40	12–30
Eurasian Woodcock	2		4	8	30	42–90	36–72
Black-tailed Godwit		2	12	36	36	36	6–12
Eurasian Curlew	6	5	7	30	38	36	8–12
Common Redshank			1¼			12–15	4–9

Table 7.1. *Prices charged for waders from the 13th to the 20th centuries in pennies (d) per bird (see Appendix 5). Prices shown for the 14th and 16th centuries are averages over the whole century. Otherwise prices are for single years. Sources Gurney (1921), Gladstone (1943), Jones (1965), Woolgar (2006).*

Great Snipe

One present day rarity of considerable interest here is the Great Snipe, *Gallinago media*. Although Brown & Grice (2005) recorded only 59 in England from 1958, this was a scarce but regular autumn passage migrant in the 19th century, mainly in the eastern counties, and particularly before 1850, when it was described as not uncommon. Its disappearance as a migrant has usually been attributed to a decline in breeding birds in northern and eastern Europe resulting from major losses of wetland habitat (e.g. Brown & Grice 2005). But Brusewitz (1969) emphasised the devastation caused by ruthless shooting and trapping at leks of this species in Sweden, providing yet another example of the impact, when combined with habitat loss, of unrestricted spring and summer persecution on the populations of such birds, a recurring theme in this book. No close season then operated in Sweden; at large leks in central Sweden in the mid-19th century, nets were stretched in all directions across the display ground. Sixty could be caught in a night, without any obvious decrease in numbers being noticeable the following night. Not surprisingly, by 1860 Great Snipe had declined more than any other Swedish game bird. Enormous numbers were also shot on autumn migration; 500 in meadows at Uppsala in 1847, for example. It was common for a hunter to shoot 60 to 70 in a single day. Similar numbers were taken in Russia near St. Petersburg in August and September in the same period (Lilford 1895). Protection came too late to have any significant effect on the Great Snipe's decline.

Nets, snares and guns

Until the 19th century, when large numbers were shot with shoulder and punt-guns, the commonest ways of taking waders was by snare and net. Shore waders were taken in flight-nets (described and illustrated on page 36). Such nets were used on the Lancashire estuaries, the Solway, around the Wash and along the Lincolnshire coast, and elsewhere in Britain and in the Netherlands and north Germany. Lorand & Atkin (1989) noted that, in Lincolnshire, they were left up all the winter and caught most birds on the darkest nights, many of them being drowned.

Amongst the shore waders, the Knot was one of the most important, and it was sufficiently highly regarded for the table to be considered worthy of being presented to visiting dignitaries by Lincolnshire corporations as gifts (Lorand & Atkin 1989). Yarrell & Saunders (1882) noted that until at least the latter part of the 17th century, Knots were regularly fattened for the table, and Sir Thomas Browne described how they were fed on corn in lighted rooms 'and when they are at their height of fatness, they begin to grow lame, and are then killed as at their prime'. Ray (1678) noted they were also fattened on white bread and milk. Besides being taken in long nets, Knots were caught in great numbers in clap-nets over painted wooden decoys (Pennant 1776) and taken at night, using lights and noise from gongs or bells to confuse the birds and allow a close enough approach to take them in hand-nets (MacPherson 1897). By the 19th century Knots were usually obtained by punt-gunners, and Yarrell (1845) noted that the London markets were well supplied throughout the winter. Some very large bags were made with punt-guns, the largest of which I have noted being 603 Knot, nine Redshank and six Dunlin at Stiffkey in Norfolk, by three punt-guns firing a salvo in 1901 (Riviere 1930).

Figure 7.1. *'Tolling' Curlew on the coast (Alan Harris).*

Shore birds generally were important quarry for professional punt-gunners. Ticehurst (1909) recorded that wildfowlers in Kent eagerly awaited the arrival of the autumn Bar-tailed Godwit flocks. Tubbs (1992) analysed the diaries of William Mudge, a wildfowler and punt-gunner on the Solent between 1897 and 1953, which showed that he shot significant numbers of Oystercatchers *Haematopus ostralegus*, Grey Plover *Pluvialis squatarola*, Lapwing, Knot, Dunlin, Black-tailed Godwit, Curlew, Redshank and Turnstone *Arenaria interpres*, most or all of which he sold.

The Curlew was usually the most expensive wader in the mediaeval and early modern periods, and Edward Topsell, writing in the 17th century, described the young as excellent eating and preferred to hens in both England and France. However, Topsell described old birds as being no better than hares (Harrison & Hoeniger 1972). Yarrell & Saunders (1882) also noted that young Curlew were eagerly sought as delicacies by the moor-men in Cornwall, and no doubt elsewhere. They also recorded that in some places 'a trained dog of red colour, as much like a Fox as possible, is employed to attract the attention of the birds and induce them to pursue him, when he entices them within shot of his master, who lies hidden in a dyke'. This was known as 'tolling' in the northeast United States and was a common method there of attracting ducks to the gun, akin to the use of a dog in duck decoys (Figure 7.1).

Plovers

As noted above, plovers, mainly Golden Plover and Lapwing, were, with Woodcock, the most numerous waders accounted for in household books of the mediaeval and early modern periods. Yalden & Albarella (2009) noted that these species were well-represented in the archaeological record, with 59% of the Golden Plover records and 68% of the Lapwings dating from the Roman, Anglo-Saxon and mediaeval periods.

Plovers were not often specifically identified in household accounts, but as they were taken or available in the markets throughout the year, the term was unlikely to embrace the Golden Plover alone. Large numbers of Golden Plover and Lapwing were taken in nets from

Figure 7.2. *The plover decoy known as a 'swipe'. The first three images show a Lapwing on the swipe, the fourth a Golden Plover. From Haverschmidt (1943).*

the mediaeval period right up to the 1940s in Britain, Ireland, the Netherlands, France and Italy. The nets and methods used were similar throughout this range, and were described by Haverschmidt (1943) as 20–24m long and 2–3m wide, mounted on poles connected through pulleys by long lines to the operator, who was concealed behind a blind. Birds were lured into the catching area by stuffed decoys, and by live decoys attached to a 'swipe', a long stick which the operator pulled to make the decoys flutter into the air, attracting birds flying overhead; call whistles were also used (Figure 7.2). As birds came in to to settle with the decoys, the net was pulled over them (Figure 7.3).

Golden Plover were the most important quarry in the Netherlands, where they were taken in large numbers. Haverschmidt (1943), for example, noted that the 168 netters operating in 1938–39 took 42,000–50,000 birds, which rose to 67,000–81,000 in 1942–43. Before 1939 the bulk of the catch was exported, especially to England.

Figure 7.3. *Netting Lapwing in Ireland. From Payne-Gallwey (1882).*

Lapwings were apparently more important in England, Ireland, France and Italy (Shrubb 2007). In his memoirs (James 1986), Ernie James described in detail the netting of Lapwings in the Fens before the Second World War. He made his own nets, which had three-inch (7.5cm) meshes and measured twenty yards long by four wide (18m by 3.5m). James said

> *At each end of the net were pulleys with springs attached to them. The springs, which were always under water, were just long poles with toes which went into the ground. They were fastened down with catch pegs – pieces of wood with a notch on to which the springs were hooked. Attached to the springs was a long line which stretched to wherever I was hiding … Before the water came onto the washes I made a small raft from a wooden frame with wire mesh stretched across it, which was covered with grass… The raft was pegged into position so that it would not move and the top of it, hopefully, was just level with the water. The net had to be wider than the raft so that some of it was anchored down by the weight of water. When the springs were released the net flew over and folded the birds inside.*

The birds were called down by decoys or whistled in, and James once took 240 in a day. He sent his birds to Leadenhall market by train, and sometimes earned £100 a week from this trade before the war, a considerable sum of money. Smith & Cornwallis (1955) noted that thousands were also taken in nets in Lincolnshire, until the use of live decoys was made illegal in 1925. Such trapping continued on a smaller scale, using stuffed decoys, until 1946, when Lapwings were finally fully protected in this area.

Shooting and egging

Both Golden Plover and Lapwing were shot in some numbers. Golden Plover were mainly taken by shore-shooters and punt-gunners, and bags of between 90 and 150 at a shot with punt-guns were recorded in Ireland (Gladstone 1930). Although found in farmland in large numbers, the species does not lend itself to organised shooting like game, although it often offers fast and sporting shots by chance.

Lilford (1895) noted that Lapwings 'may easily be decoyed within gun-shot range by means of 'stales' or tethered birds' but remarked that he did not often shoot them, finding them less palatable than Golden Plover. Yarrell (1845), however, regarded them as excellent for the table. Newton (1896) observed that 'the bird, wary and wild at other times of the year, in the breeding season becomes easily approachable, and is (or used to be) shot down in enormous numbers to be sold in the markets for "Golden Plover". Its growing scarcity as a species was consequently very perceptible'. Many were still shot on the Ayrshire coast in the 1920s (Paton & Pike 1929) and Booth (1881–87) recorded that one gunner shot 2,400 Lapwings in three months on the upper Forth in the late 19th century. Booth also noted that such persecution had caused them to largely desert Pevensey Levels in Sussex. Large numbers were shot by punt-gunners on the fenland Washes (Cocker & Mabey 2005).

They are still shot in France, Spain, Italy and Greece. The largest numbers are shot in France, where Trolliet (2003) estimated that 1.6 million Lapwings were shot in 1983–84. Since then fewer hunters and a shorter shooting season have resulted in a decline in the numbers shot, and Trolliet (2003) estimated that 436,000 were killed in 1998–99. Trolliet quoted figures of possibly 100,000 taken in Greece, up to 250,000 in Italy and an unknown number in Spain, giving an overall estimate, including birds crippled but lost, of around one million birds still being taken each year.

The Lapwing was also an important egg species. Cott (1953) estimated the crop taken annually at 100,000–1,000,000 before the 20th century. But this may have been an underestimate, for Newton (1896) noted that 800,000 eggs were imported into London annually from Friesland in the Netherlands alone in the 1870s. Eggs were collected widely in Britain, but the statistics of this trade are rather fragmentary. Even so, they show that thousands of eggs were gathered for sale in the 19th century, and such statistics ignore the widespread casual collection of eggs for food by farm workers, fishermen and so forth. Lapwing eggs were always valuable, and authors from the late 18th century quoted prices of three to ten shillings per dozen (£15–£50 in today's money). As Lapwings declined and egging was restricted prices rose, and early-taken eggs may have commanded as much as 18 shillings per dozen (£84 today) or more in the 1890s (Cott 1953).

A fuller account of the exploitation of Lapwings is given in Shrubb (2007).

The Dotterel

In Britain, the Dotterel *Charadrius morinellus* mainly breeds today in the Highlands of Scotland, where recent estimates indicate around 630 breeding males (Whitfield, 2002). It occurs in England and Wales mainly as a spring passage migrant, but breeds in very small numbers in the Lake District. In the past Dotterel were much sought after as table birds and, in the 19th century, shot in large numbers for their feathers, used for tying fishing flies.

Dotterel have not been recorded in the archaeological record in Britain (Yalden & Albarella 2009). Surprisingly for a species so highly valued for its edible qualities, they were also unrecorded in the sources used to compile Table 1.3 (p. 20) before the 16th century. Nor does it appear in Jones's (1965) price lists before the same date, and it is also absent from the Neville feast of 1465 (Table 1.5, p. 22). One possible explanation is that, as plovers were often not specifically identified in early household accounts, Dotterel were included under that heading. Montagu (1833) remarked that it was often confused with Golden Plover in the past, and Bucknill (1900) gave at least one record confirming this. But it is also probable that the status of Dotterel in Britain changed markedly with the onset of the Little Ice Age from the mid-14th century, which was at its most severe from the late 16th to the early 18th centuries (Fagan 2000). This may have caused a southerly shift in Dotterel breeding distributions, with a concomitant increase in breeding birds and migrants in Britain. Records to support this suggestion are lacking, but that it is plausible is indicated by the history of the species in the 19th and 20th centuries (see Nethersole-Thompson 1973, Burton 1995, Hagemeijer & Blair 1997). Harvie-Brown (1906) gave an interesting example of this. He noted a sharp increase in Tayside, where the species had long been declining, in the unusually cold spring and summer weather of 1902–04. He considered the increase to be was the result of a southerly shift in the Dotterel's breeding distribution in response to the weather.

In the 16th century the Dotterel was among the species Lord Lisle (see p. 24) was sending to England from Calais, but he was asked to stop sending them alive, as they were then being obtained from Lincolnshire (Bourne 1999b). It was clearly a common spring passage migrant in Britain by this time, with regular stopover points at sites on the chalk uplands of Cambridgeshire, the Brecks of Suffolk and Norfolk, the Lincolnshire Wolds and coastal marshes, the Yorkshire Wolds, coastal fields, marshes and upland moors in Cumberland and Northumberland, and the southern uplands of Scotland, particularly in Berwickshire. They were also regular in some numbers on the Lancashire mosses and coastal marshes, and on the chalk uplands of Sussex, Wiltshire, Berkshire and Buckinghamshire. They were apparently much rarer in western districts of Britain.

The spring passage appears always to have occurred in a fairly concentrated period, usually from about the third week in April to mid to late May but there are sufficient early 19th century records, particularly in eastern England, to suggest that passage may then have started around mid-March. However, peak movement was always concentrated in May.

Spring passage and persecution

Although detailed counts are lacking, it is clear that before the 1850s, the Dotterel was a numerous spring migrant, particularly in northern England and the Scottish borders. In Lancashire, for example, Mitchell (1885) noted that it was then so numerous on sites such as Pilling Moss that hundreds were offered for sale in Preston market in May. The shooting of spring migrants continued in Lancashire until 1920 (Oakes 1953). Ratcliffe (1973) noted that trips of up to 200 birds were regular in spring on the coastal salt marshes, lower moors and high fells of Lakeland. Many were shot, particularly for their feathers for fishing flies.

Similarly in Yorkshire, Nelson (1907) noted that the 'breast feathers of this bird were formerly, and still are, in great request by fly fishers, and such was the demand for them in comparatively recent times that, from the Holderness coast right up to the high grounds about Bempton and Speeton, the shooting of Dotterel was a regular occupation in spring'.

He went on to remark that some gunners reckoned to have shot 50 brace in a season. Larger numbers were procured in some areas on the Wolds, with up to 42 brace in a day shot at Ganton, Sherburn and Knapton. In the mid-19th century the bird was also numerous on the Hambleton Hills, where J. H. Phillips (in Nelson 1907) said he put up hundreds on the moors. It was also sought regularly on the Wensleydale moors, again for feathers for fishing flies. Many of the gunners who visited these Yorkshire hills in pursuit of Dotterel were gamekeepers, for whom Dotterel-shooting appeared to be an annual perk – and a valuable one, with prices of at least one shilling (*c.* £5 today) per bird. MacPherson (1892) noted that the feathers alone were worth half that for fly-tying. The Strickland family built the Dotterel Inn at Reighton on the Wolds, apparently for the accommodation of gamekeepers coming from all over the neighbourhood for Dotterel shooting (Nelson 1907).

The situation was similar in Berwickshire, where Muirhead (1895) noted that up to 50 years before, great numbers of Dotterel had visited the high ground in 16 parishes along the Lammermuir Hills. He gave less detail about the shooting of Dotterel but noted that James Purves, gamekeeper at Foulden, sometimes bagged 10–15 brace in a day on the Lamberton farms towards mid-May, when great flocks frequented newly sown fields on the edge of the moor. Muirhead noted that at this time, Dotterel were 'in great request by sportsmen in the county and visitors from a distance, not only on account of it being a very dainty dish … but also because it came in at a time when game was out of season'.

Other upland areas in northeast England, from the Wolds and heaths of Lincolnshire (and its coastal marshes) to the uplands of Durham and Northumberland, also attracted large flocks in this period. These were similarly persecuted. Large spring movements also occurred in parts of Europe, and were greeted in the same way. In Denmark, for example, Meltofte (1993) recorded that 5,200 were shot at three sites in Ringkoping Amt in the spring of 1884.

There is no data by which to estimate what proportion of these spring passage flocks represented the British breeding population, although Ratcliffe (1973) inferred that the flocks in Cumberland probably were Lakeland breeding stock. But the numbers involved suggest that both the British population and birds from Scandinavia were involved. While it is unclear how much duplication there was in the records, with flocks moving on and being recorded at other sites, the records available indicate that before the mid-19th century Dotterel must have passed through northern England and the Scottish Borders in their hundreds in the spring, probably in thousands. Furthermore, in eastern Scotland north of the Firth of Forth, the series of *Vertebrate Faunas* of the 19th and early 20th centuries, whilst recording spring passage north to Caithness, implied that these birds were local breeding stock; what detail on spring-passage numbers they gave indicate much smaller numbers than further south.

In southern and eastern England, Dotterel were regular spring migrants on the chalk uplands of Cambridgeshire, where large numbers were caught on the Gog and Magog Hills, and the heaths, Brecks and fens of Suffolk and northwest Norfolk into the early 19th century. These areas, particularly the Cambridgeshire chalk, regularly attract Dotterel in the spring today (Bircham 1989). They appear to have been even more numerous on the Wessex chalk in the early 19th century. Smith (1887) noted that they were then annual spring and autumn migrants on the Wiltshire Downs, resorting to the new sown corn and fallows. He recorded that 'they generally rested but a few days amongst us, but during that period they were often so numerous that sportsmen now alive have killed from 40–50'.

Hunting Dotterel in earlier times

The traditional mode of hunting Dotterel in East Anglia in the past was by netting them (Stevenson 1870). James VI/I in 1610 flew his Sparrowhawk *Accipiter nisus* over Dotterels at Thetford so that they could be taken, probably with a net (Newton 1879). Newton remarked that he knew of no other record of Dotterel being taken in this way, but it was a method used for taking partridges and Quail with a setting-net, so it is quite possible that it was applied more frequently to Dotterel. Ray (1678) described the usual way of netting them in Norfolk as

> ... *to catch Dotterels six or seven persons usually go in company. When they have found the birds, they set their net in an advantageous place; and each of them holding a stone in either hand get behind the birds, and striking their stones often one against another rouse them, which are naturally very sluggish, and so by degrees coup them, and drive them into the net.*

They were also taken at night by dazzling them with a lantern to allow a close approach, then taking them in a hand-net.

With the exception of Smith (1887), whose account implied that they were as numerous in autumn as in spring, all the main accounts for the 19th century stress that migrant Dotterel were always far fewer in the autumn in Britain; modern records indicate that this remains so. Stevenson (1870), for example, remarked that 'during the last 16 or 17 years I have never seen a single Dotterel in autumn, either in our poulterers or birdstuffers' shops', although a few still visited the Breckland warrens in August. Most authorities noted that the majority of autumn migrants were young birds. The paucity of autumn migration might suggest that a different route to that taken in spring was followed in autumn, but evidence is lacking. On the other hand, the pattern of movement may have differed, which was also suggested by early 19th century records from Belgium, where Deby (1846) noted that Dotterel were seen annually in the first two months of autumn but that he had no spring records.

Decline

After the 1850s the numbers of Dotterel on spring migration in Britain declined steadily. The species was increasingly sought on its British breeding grounds, for specimens and eggs for collectors as well as for the traditional uses as table birds and for fly-tying feathers. It is noteworthy that, although the species was theoretically protected under the Protection of Birds Act of 1896, none of the Scottish counties in the core breeding area were covered by protection orders of the Secretary of State for Scotland (Marchant & Watkins 1897), so it is doubtful that any effective or even theoretical legal protection was provided for the bird. By the early 20th century Dotterels had become distinctly scarce on spring passage. Lack (1934), for example, noted only six records for Cambridgeshire between 1900 and 1934, and Bircham (1989) only two more by 1949, after which spring migrants started to appear regularly once again. In Yorkshire Chislett (1953) could list just 12 records between 1909 and 1953, and on the Wessex chalk Dotterels had become rarities by the end of the 19th century (Smith 1887, Standley *et al.* 1996) as they had in Sussex (Walpole-Bond 1938).

Most 19th-century authors had no doubt that the scale of spring shooting was the main cause of this decline. Stevenson (1870) pointed out that spring shooting was simply destroying potential breeding stock. Ratcliffe (1973), however, observed that the Lakeland population continued to decline despite a decline in persecution, indicating that some other factor was also at work; he suggested that the climatic amelioration of the late 19th and early 20th centuries had caused a northerly shift in breeding range. The decline of spring migrants went on all over Britain, and in northwest Europe. It seems likely that the recorded scale of spring persecution affected breeding populations in these areas. Furthermore, in an analysis of 150 years records in Finland, Saari (1995) concluded that there had been a decline of 90% or more in the breeding population, mainly in the 19th and early 20th centuries, and that this had probably been due to severe overhunting. Thus it seems probable that the recent increase in the Scottish breeding population in a period of continued climatic amelioration represents recovery of this species from the effects of excessive human predation, although the longer-term trend may be to see breeding ranges shift further north.

The Ruff

The Ruff *Philomachus pugnax* formerly bred in fen habitats in East Anglia, Lincolnshire and southern Yorkshire, at Prestwick Cars in Northumberland, Martin Mere in Lancashire and, according to Montagu (1833), in the Bridgewater area of the Somerset Levels. Early accounts confine themselves to discussing the population in general terms. Lubbock (1845), for example, described Ruff as 'swarming' in the Broadland marshes in Norfolk, and Ellis (1965) speaks of large colonies formerly breeding in these marshes. Sir Thomas Browne described them as 'abounding' in the Fens in the 17th century (Stevenson 1870); Pennant (1776) noted that the Fen fowlers could take 40–50 dozen each in a season, and recorded one taking 72 birds in a day, while John Clare, in the early 19th century, described Ruff as arriving in 'great droves' in the Fens in the spring (Robinson & Fitter 1982). William Farrer's father recalled seeing large baskets full of Ruffs and reeves netted in the Fens coming into Cambridge market in the 1840s (Lack 1934). In Yorkshire they were still described as common in many fenland localities in the early 19th century (Nelson 1907).

Despite this abundance, Ruffs occur in the archaeological record and household accounts for the mediaeval period surprisingly infrequently. Nor was the bird common in the market lists published by Jones (1965). Bourne (1999b) suggested that the term *rees*, normally interpreted as Ruff or reeve, should actually be interpreted as small shore birds generally, which seems unlikely in view of the value (2d each) placed on them in the Northumberland household book, for example (Nelson 1907). Pennant (1776) offered another interpretation, arguing that the term *rey* or *ree* meant the Corncrake (see p. 113).

Trapping and fattening Ruffs

Ruff were usually fattened for the table, and there are indications in the early literature that this was a professional trade, rather than a process carried on in the household as with birds such as herons or plovers. By the early 19th century, Ruffs had declined considerably, largely as a result of land drainage, and in Lincolnshire, formerly an important centre for catching Ruff, Montagu (1833) recorded that the Ruff-catching trade was 'confined to a very few persons and, at present, scarcely repays their trouble and expense of nets'. The birds were sold to specialist fatteners, who paid the fowlers that caught them around 10 shillings per dozen, whilst selling them on for three or four times that amount.

Montagu visited two such specialists, a Mr Weeks of Cowbit and Mr Towns of Spalding. The birds were fattened in large rooms, where the males distributed themselves as though at their leks, each with its own stand. Males and females were fattened together and were fed on bread and milk, hempseed or boiled wheat. Sugar was added to speed the process if desired (Pennant 1776). On his visit to Weeks, Montagu was shown seven dozen males and a dozen females in one room, the males in full plumage 'and not two alike'. Yarrell & Saunders (1884) suggested that comparatively few Ruff were taken in the spring as older birds did not fatten well, and that September was the principal season for the trade, but this is not borne out by other authorities.

Montagu's visit to Mr Towns gave some insight into how the trade in Ruff operated. Towns reported that his family had been engaged in it for a century, and had supplied George II and many noble families in the kingdom in that time. When the Marquis of Townsend was Lord Lieutenant of Ireland, Towns undertook

> … to take some Ruffs to that country; and actually set off with 27 dozen, from Lincolnshire; left seven dozen at the Duke of Devonshire's at Chatsworth; continued his route across the Kingdom to Holyhead; and delivered 17 dozen alive in Dublin; having lost only three dozen in so long a journey, confined and greatly crowded as they were in baskets, which were carried upon two horses (Montagu 1833).

In Lincolnshire, Ruffs were taken in nets, as they were elsewhere in the Fens and in Yorkshire. According to Montagu, they were taken on the leks with a single clap-net about 17 feet (5m) long, set before daybreak with a number of stuffed decoys. Most were taken at dawn at the first pull of the net. Single birds were taken subsequently, enticed by decoys. The birds were at their least shy when the reeves began to lay; one fowler assured Montagu that he could then take every bird on the fen. Pennant's (1776) account indicates that this was the common method of taking Ruff in the Fens, but in Yorkshire they were caught using a rather different form of net, which was mounted on poles set at an angle of 45° and dropped on birds enticed beneath by decoys. Montagu noted that the methods used to catch Ruff differed from spring to autumn, but he did not elucidate. As lekking birds were unavailable in Autumn, they were probably netted over decoys at that time.

Apart from in the Fens, Ruff were not generally netted in Norfolk; rather, they were caught on the lek with horse-hair snares. No information about the numbers taken is now available, except that Lubbock (1845) noted that one old fowler told him that he had only once 'caught six couple' in a morning. Lubbock also reported that many Ruff eggs were taken in the Broadland marshes and sold as Lapwing eggs. He made the point that, unlike Lapwing, Ruff were less inclined to re-lay; such robbery harmed these birds more.

Population decline

Ruff declined very sharply in the first half of the 19th century, and ceased breeding altogether, except sporadically in Lincolnshire and Norfolk, by the middle of the century, largely as the result of the drainage and conversion of their breeding grounds to arable agriculture. But exploitation undoubtedly contributed and, as the species became scarcer, its eggs were increasingly sought by collectors for their cabinets and specimens, particularly males in breeding plumage, for their display cases. Stevenson (1870) noted that such specimens were 'exactly the creature which all bird preservers eagerly snap up; being purchased not merely by the naturalist, but by any one desiring a pretty object in a glass case'.

Although the market ceased to be supplied from England on any scale by the mid-19th century, Ruff were offered for sale in the London markets into the 1920s (Gladstone 1943). Both Harting (1901) and Lilford (1895) noted that substantial numbers were imported live into London in April from the Netherlands, a trade that had been in existence since at least the early 17th century (Haverschmidt 1963), and MacPherson (1897) noted that Ruffs were also imported from Denmark. They were fattened on arrival on bread, milk and boiled grain and found a ready sale; Lilford (1895) remarked that the proportion of males to females was about 10:1. Harting noted that such supplies had virtually ceased by the early 20th century, but some imports certainly continued into the 1920s, and recipe books of the 1930s still gave instructions for preparing Ruff for the table.

The Black-tailed Godwit

In Britain, the history of this species, *Limosa limosa,* which was known as the *Yarwhelp, Yarwip* or *Yelper* in the Fens, is similar to that of the Ruff. Godwits bred in the same habitats, were similarly exploited as table birds and disappeared as breeding birds by around 1835 as a result of over-exploitation combined with drainage. The Black-tailed Godwit was less widely distributed than the Ruff, being largely confined to the Fens, the Broadland marshes in Norfolk and the Hatfield moors in Yorkshire. Early accounts suggest that it was never particularly abundant.

As a table bird, Yarrell & Saunders (1884) noted that Black-tailed Godwits were considered a luxury in the 16th century, and Dr Muffet, writing in the same century, noted that 'a fat Godwit is so fine and light a meat, that noblemen, yea, and merchants too, by your leave, stick not to buy them at four nobles a dozen' (a noble was worth 6/8d). Not everyone thought the same. Isaac Casaubon in 1613 (quoted in Darby 1934) said that 'the flesh when cooked is dark as is that of marsh birds. I ate it at the Lord Bishop's table and did not think highly of it'. In the 18th century, fattened godwits sold for between 2/6d and 5/- each (Pennant 1776). William Turner, writing in 1544, remarked that in captivity it fed on wheat 'just as our pigeons do' (Evans 1903). Gladstone (1943) showed that they were sold in London into the 1920s, when they were priced at 6d to a shilling each (80p to £1.65 today).

There is little in the early literature about the way in which this species was hunted. Pennant (1776) noted that they were netted over decoys as Ruffs were; none of the early avifaunas seem to have anything significant to add to this.

After it ceased to breed it became an increasingly rare passage migrant, much collected for display cases. Stevenson (1870) noted that one reason for its increasing scarcity as a migrant on the east coast was that this formerly abundant breeder in the Netherlands had 'like the Purple Heron, Spoonbill and Little Bittern been so destroyed there of late years, that it has become comparatively rare'. Black-tailed Godwits (and Ruff) had been netted in the Netherlands and exported to England since at least the early 17th century, and probably before. In 1628 their export from Friesland was forbidden, except for 500 birds (and a similar number of Ruff) twice a year to the king of England. This ordinance was repeated in 1640 (Haverschmidt 1963).

The eggs of this species were also collected for food in large numbers in the Netherlands, particularly in Friesland. Again, since the 17th century or before, large numbers of eggs were exported to England, where they were much in demand. Haverschmidt (1963) noted that the numbers taken must have run into many thousands, although exact figures are lacking. However, the principal game dealer of Friesland sometimes had 2,000 eggs for sale at the end of the season. This trade ceased as recently as 1956.

The Snipe

Yalden & Albarella (2009) recorded the Common Snipe *Gallinago gallinago* as well-represented in the archaeological record for Britain, with 68% of the records dating from Roman times onwards. It was the fourth most frequently recorded wader species in the lists upon which Table 1.3 (p. 20) is based. Large numbers of Snipe were sold in the London markets in the mid-19th century, Mayhew (1861–62) recording an average total of 107,000 annually through Leadenhall and Newgate markets for example.

Throughout Europe, Snipe were taken in nets and snares, often in specially prepared trapping areas made into attractive feeding sites; the Dutch, for example, spread pig manure on the site, while the Poles laid boards on the grass to encourage earthworms to work on the surface (MacPherson 1897). In Lancashire lines of snares, known as pantles, were set along

Figure 7.4. *A Lancashire Snipe-pantle. From Mitchell (1885).*

prepared runs in the grass. Mitchell (1885) gave a detailed description of these devices, which comprised pairs of horsehair snares set about 7cm apart, along a heavy horsehair line 12m long set just above the ground (Figure 7.4). The Poles used similar snaring devices, but the Dutch preferred clap-nets. The netting and snaring of Snipe died out to a large extent in Britain in the 19th century with enclosure and the increasingly close preservation of game, which discouraged such activities. Snaring continued for longer in Ireland, and Payne-Gallwey (1882) described in detail the use of springes for Snipe and Woodcock there (see Figure 7.5, p. 130).

Snipe as gamebirds in Britain

Snipe were important for hunting, particularly in Ireland and the Western Isles of Scotland. Some very large bags were recorded in the 19th and early 20th centuries, 1,108 on Tiree in nine days in October–November 1906, for example, and 1,292 there in 11 days in the same period in 1908. One man shooting on Orkney in 1908–09 killed 2,344 and, in Ireland in the season of 1845–46 on Lord Sligo's estate in Mayo, one gamekeeper shot 3,300, and 3,330 the following season. In 1924–25 2,009 were shot at Frenchpark, Roscommon (Gladstone 1930). It was not only estates that produced such bags. Gladstone recorded that Patrick Halloran, a professional wildfowler of Kilker in County Clare, killed more than 2,000 Snipe in 1879–80, and 1,376 the next season. Altogether, in 50 seasons he shot more than 40,000 Snipe. Throughout Britain and Ireland records by Gladstone (1930) and by Payne-Gallwey (1882) show that bags of between 100 and 300 in a day were commonplace, and sometimes achieved by a single gun.

Such bags give a clear indication of just how abundant Snipe must have been in the 19th and early 20th centuries. They are no longer possible today with the scale of wetland drainage during and since the Second World War (see Shrubb 2003), and Tapper (1992) showed a steep decline in the numbers shot since the 1930s.

An interesting question arising from these bag records was how many Jack Snipe *Lymnocryptes minimus* were involved. Few records were kept, but Payne-Gallwey (1882) noted that they were becoming rarer annually in Ireland, and quoted a Colonel Peyton that a proportion of two Jack Snipe to 25 Common Snipe was the average in western Ireland after 1860, which represented a decline of 20% since the early 19th century. Gladstone (1930) recorded that the bag of 2,344 Snipe in Orkney noted above was accompanied by a total of 411 Jack Snipe (15% of the total), and also recorded that of 8,679 snipe shot by Mr Charles Bouck between 1893 and 1929 in various parts of Britain, 1,917 (22%) were Jack Snipe. In Wales, S. W. White analysed his snipe shooting records covering 60 years in Breconshire, and found that the proportion of Jack Snipe shot was one to three Common Snipe (Ingram & Salmon 1957). If the estimate of one Jack Snipe shot for every eight Common Snipe given today is correct, the impression of scarcity given by field observations of this skulking species is much exaggerated (Harradine 1986).

The Woodcock

Yalden & Albarella showed that the Woodcock *Scolopax rusticola* was the most frequently recorded wader in the archaeological record in Britain; 30% of the records were for the

Roman period and 52% for the mediaeval period or later. Household accounts and market lists show a similar status, with Woodcock the second most frequent and second most abundant wader species recorded in Tables 1.3 (p. 20) and 1.4 (p. 21), behind the plovers, a category that actually included more than one species. This status undoubtedly reflected the Woodcock's importance as an article of food, a subject discussed in some detail by Thirsk (2007); a favourite way of preparing and preserving them was in Woodcock pies and puddings.

In this early period Woodcocks were mainly known as abundant passage migrants and winter visitors. For household accounts that include date information, by far the bulk of the birds accounted for were purchased during the last three months of the year, particularly October and November. The species' status in the breeding season at this period is impossible to assess, but household and market records show that it was certainly present. For example, the lists in Jones (1965) show it being offered for sale during April to July, particularly in the latter month, a date that could hardly apply to migrants, while the Shuttleworth family of Lancashire was sending Woodcock puddings to someone in London in May 1587 (Thirsk 2007). The 400 listed in the provisions for the Neville feast of September 1465 may also have been largely home-grown, as autumn migration would have barely started by this time.

By the 19th century, the early avifaunas indicate that it was a widespread but uncommon breeding bird in Britain, but it increased significantly as the century progressed. Yarrell & Saunders (1884) attributed this to the increase of plantations during the century, particularly as pheasant coverts, although Booth (1881–87) suggested that the bird was much overlooked before 1850.

Hunting Woodcock

The Woodcock was an important quarry for the game shooter and from the end of the 18th century was protected in some areas for this; before this date, birds were mainly taken in nets and snares. The usual way of netting Woodcock was in a cock-shoot or cock-road. This was an angle cut into woodland (see Figure 8 in Rackham 1990), or a glade or ride specially cleared in a wood, which Woodcock used as flight paths on leaving the wood to feed at dusk (known as cock-shoot time). A net suspended on pulleys was stretched across this route and caught the birds as they flew through. Similar nets were also sited along hedgerows that were known to be followed as flight paths. Such nets were used throughout much of western Europe, and more than one cock-shoot might be sited in a single wood; Newton (1896) quotes one example of a wood having 13. Ten or twelve such nets were also licensed for use on Heligoland, an island of around 260ha (Gatke 1895).

Considerable numbers of Woodcock were taken in these nets. Writing of Pembrokeshire in 1603, George Owen commented on the number of Woodcock found there from Michaelmas to Christmas, and noted that 'it is no strange thing to take a hundred or six score [120] in one wood in 24 hours' (Mathew 1895). MacPherson (1897) recorded that up to 30 dozen Woodcock a week were sent from Devon to London in the 18th century on the Exeter stagecoach, all netted in cock-shoots, and that some cock-shoots in Picardy yielded 700–800 birds annually.

The usual method of snaring Woodcock was with a springe, a device in which a horse-hair snare was attached to a pliable rod stuck in the ground and bent over to catch on a

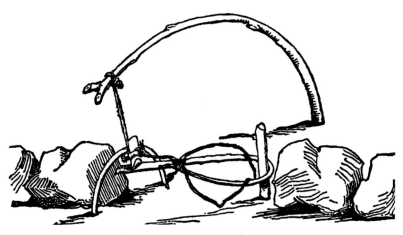

Figure 7.5. *A Lake District Woodcock-springe. From MacPherson (1892).*

crude trigger. When a bird was caught by the snare the trigger was released and the springe flew up, carrying the bird with it and strangling it. These devices were widely used in Ireland (for Snipe and wildfowl as well as Woodcock) and in northwest England (Payne-Galwey 1882, MacPherson 1897). MacPherson noted that in northwest England it was customary to set them in gaps along little low walls or hedges specially constructed to encourage the birds to run along them and into the snares, Woodcock being disinclined to hop over such obstacles (Figure 7.5). The numbers taken were considerable. In the household accounts of the Howard family at Naworth, for example, 214 were paid for between September 26th 1618 and February 7th 1619, and in the last three months of 1621, 154 were purchased (MacPherson 1892). Pennant, in his *Tour of Scotland*, saw in the neighbourhood of Windemere in 1772

> *on the plain parts of these hills numbers of springes for Woodcocks, laid between tufts of heather, with avenues of small stones on each side to direct these foolish birds into the snares for they will not hop over the pebbles. Multitudes are taken in this manner in the open weather and sold on the spot for sixteen or twenty pence a couple (about 20 years ago at sixpence or seven pence) and sent to the all-devouring capital by the Kendal stage.*

Despite the rather fragmentary nature of the records available there is little reason to doubt that Woodcock were extremely numerous winter visitors and passage migrants in the past. They were abundant enough in the late 17th century for John Houghton, a reputable journalist with a strong interest in food supplies, to float the idea of exporting thousands of Woodcock pies to help feed the American colonies (Thirsk 2007). That such a suggestion was thought feasible sheds an intriguing light on the bird's numerical status. Some 19th-century authors, Stevenson (1870) and Mitchell (1885), for example, suggested that the number of passage migrants and winter visitors declined in the second half of the century. But this does not seem to have been the general opinion, and such declines may have resulted from local changes in habitat, which Mitchell certainly inferred. Changes in weather patterns associated with the climatic amelioration in the second half of the century may also have changed patterns of movement. This was certainly the case with Woodcock and many other migrants on Heligoland (Gatke 1895).

Woodcock bags

Although enclosure and the increasingly close preservation of woodlands for pheasants in the 19th century greatly reduced the opportunities of taking Woodcock in these traditional ways, large numbers were shot in the 19th and early 20th centuries, particularly in Ireland, western Scotland and southwest England. Some estates, particularly in Ireland, appeared to specialise in Woodcock-shooting, and Payne-Galwey (1882) noted that a good yearly average bag on a favoured estate in Ireland was 300–400 birds; double that number could be taken in good seasons, for numbers always fluctuated with the severity of the winter further north and east. The severe winter of 1880–81 saw exceptional numbers in Ireland, and huge numbers were taken. One game dealer in Tralee, for example, handled 2,021 Woodcock between October and February, shot by the peasantry and professional fowlers over much of County Kerry. Another dealer in County Clare forwarded 3,000 Woodcock in three weeks over the course of that winter to Dublin and London. Payne-Galwey (1882) examined many birds shot that winter, and observed that they were almost without exception in excellent condition. He often found them feeding on mudflats at the coast.

Equally large bags were taken on some estates in western Scotland, particularly on Islay, where seasonal totals of 400–500 birds were not unusual in the early 20th century. On the Ardimersay shoot there, 5,066 Woodcock were shot in nine seasons between 1848 and 1858, and 3,266 between 1905 and 1914 (Gladstone 1930).

Numbers shot in England seem never to have been so large as those in Ireland or Western Scotland, perhaps reflecting a drier and colder climate. More probably, however, the greater importance of traditional game-shooting in England limited opportunities for shooting Woodcock. The record English bag is 106 shot at Lanarth in Cornwall, an estate where particular care was taken to attract Woodcock, in 1920. Kelsall & Munn (1905) recorded 110 being shot in three days in Parkhurst Forest in the 1840s. Otherwise, Gladstone (1930) found only one other bag-record of more than 100 being shot in England in a day, of 105 shot at Swanton Wood near Melton Constable, Norfolk, in either 1860 or 1872.

CHAPTER 8
Seabirds

For centuries, seabird colonies formed an important, often essential, source of protein, feathers and oil to coastal communities, particularly on isolated islands. Such colonies had obvious advantages as a food source, being densely populated, with the birds themselves being meaty with good-sized eggs. Seabird colonies were also vulnerable to simple fowling techniques, although the level of skill and courage needed to harvest some species should never be underestimated. An indication of their importance in some parts of Britain is reflected in the fact that domestic fowl were late in reaching the north and west of Scotland, as the need for eggs, meat and feathers supplied elsewhere by domestic poultry was met by abundant seabirds (Serjeantson 1988).

The exploitation of seabirds in northwest Europe

Serjeantson (2001) remarked that the failure of early farming communities to develop conservation measures for seabirds was likely to have contributed to the decline of species such as the Great Auk around British shores in prehistoric times. But by historic times, most communities dependent on seabird colonies around the eastern seaboard of the Atlantic had evolved measures to prevent over-exploitation. Such measures required some form of governance to enforce. For eastern Atlantic seabirds, this took the form of communal ownership and control, state ownership (in lowland Scotland and England usually through a feudal lord), or by the private ownership of individuals. Seabird colonies were valuable properties, a source of revenue and tax income as well as food and tradeable commodities, and each of these mechanisms exercised control over access to colonies and the nature and scale of exploitation (Serjeantson 2001). For example, the important Northern Gannet *Morus bassana* colony on the Bass Rock, which was a fortress of the Scottish Crown, was held by the Bishop of St Andrews, and granted to Robert Lauder by a charter of 1316. The Lauder family held it until they sold it in 1621 (Gurney 1913). When Pennant visited the Rock in 1769 he found all its birds strictly protected, with no one being allowed to shoot at the Gannets, 'the place being farmed principally on account of the profit arising from the sale of the young of these birds, and of the Kittiwake' (Gurney 1913).

In the Hebrides, the rights to seabird colonies on uninhabited islands often belonged to the tenants of the nearest inhabited areas; rights on the Flannan Isles, for example, went to tenants on the west coast of Lewis. Martin (1703) noted that the men of western Lewis visited these islands once each year in summer in the late 17th century 'and there make a great purchase of fowls, eggs, down, feathers and quills'. Rights to the gannetry on Sula Sgeir belonged to the people of Ness, also on Lewis (who still harvest Gannets there). The great puffinry on the Shiant Islands in the Minch was exploited by the men of Leumrabhaigh on Lewis (Mackenzie 1949).

The main beneficiaries of seabird colonies in Orkney were the proprietors of the land, having seabird rocks divided among them as they held their lands. Similar arrangements operated in most of Shetland, where the small uninhabited holms often bore sizeable seabird colonies. Such sites were protected by severe laws against trespass, but on three important Shetland seabird sites, Foula, Fair Isle and Noss, the value of the seabird colonies went to the local population (Shaw 1980, Serjeantson 2001).

On St Kilda, where the human population depended very largely on seabirds for food, fowling rights were rigorously guarded. Some bird-cliffs were divided between the islanders, and some were common property, and in both cases all cliff- and bird-shares were determined by the amount of land a man held. Allocations were exchanged every three years (later annually) to ensure fairness. Where rocks were common property the harvest was divided out among the community, so that all had a fair share. In the Faroe Islands some bird-cliffs were held by specific individuals or groups of individuals, and some were common property. Cliff fowling rights were determined by the amount of land owned in the original infield of the settlement, and land later enclosed carried no rights. There were also complex rules concerning how many birds an individual could keep and how many had to be distributed to the community (Baldwin 1974). These arrangements did not apply to the Northern Fulmar *Fulmarus glacialis* as they were in force long before that species colonised the islands (Williamson 1945). Nor did they apply to species of minor importance, such as the Black-legged Kittiwake *Rissa tridactyla*, which could be taken by anyone with the permission of the landlord.

But the state also exercised some control over the conservation of seabirds in the Faroes, where breeding cliffs were protected throughout from March 1st to August 15th. During that period no gun was permitted to be fired at sea within two miles of a colony, or on land within one mile (Fielden 1872). As Williamson (1948) noted, many Common Guillemots *Uria aalge* were shot at sea in the spring by villagers with no fowling cliffs, so the wisdom of this prohibition is clear. Fielden remarked on the tameness of the auks within this protected zone and their wariness outside it. He also noted that European Shag *Phalacrocorax aristotelis* colonies were protected throughout the year, and gannetries from January 25th to October 25th.

Numbers of seabirds taken in northwest Europe

Early historic information on the numbers of seabirds taken in northwest Europe is scant and often unreliable. But Table 8.1 provides some figures, mainly from the later 19th and early 20th centuries, for four of the most important species from three important seabird fowling sites. The harvest mainly comprised young birds at the point of leaving the colonies, or when they first took to the water; many young guillemots, for example, were rounded up and killed in the water from boats before they swam out to sea. Relatively few adults were taken, except that large numbers of adult or fully grown non-breeding Atlantic Puffins *Fratercula arctica* were taken with the *fleygastong* (pole-mounted triangular net; see p. 142) in the Faroes and Iceland. Fulmars were also fleyged. Large numbers of eggs were also taken from species such as guillemots, which lay replacements.

a) St Kilda

Fulmar	Certainly breeding end of 17th century. The inhabitants preferred it as food to all other birds, save possibly Gannets; *c.* 12,000 young taken 1841. Mean taken annually 1840–1910 (based on six years) 10,793, mean 1874–1929 (based on 12 years) 9,175, a fairly consistent 40–50% of young produced.
Gannet	Young probably taken since at least mid-16th century. At least 800 taken Stac an Armin 1696; 5,000–7,000 stated to have been taken Stac Lee 1697; up to the early 1840s never more than 5,000 young taken in a year, and 2,000–3,000 adults; 3,200 young taken 1895. By 1902 only 300 taken; 1910 600 adults taken but no young.
Common Guillemot	Large numbers of eggs taken but no reliable estimate of birds. Adults not thought good eating, except on first arrival in March and April. Mostly used as a source of feathers and to manure the fields.
Puffin	More killed than any other species. In 1840 chief article of food but subsequently taken mainly for feathers. At least 89,600 killed in 1876. But many fewer by end of 19th century as demand for feathers fell; still being taken at time of evacuation of island in 1930.

b) Iceland

Fulmar	Up to 150,000 killed in some years. On the Westmann Islands 20,000 taken in 1821, 20,000–30,000 in *c.*1852 and, in the 1890s, when 30,000 usually taken in a good year, 24,229 in 1898. On Grimsey 3,000–4,000 taken in some years. In Myrdalsfjall 20,000–25,000 taken annually by 1936. Psittacosis found in the population in the 1930s and killing of young stopped. This ban led to egging on the Westmann Islands, which may have reduced the population.
Gannet	On the Westmann Islands 400–500 young taken in the 1860s, 562–662 in 1898 and an annual mean of 508 during 1913–1930. On Eldey 4,100 young taken per year around 1908, and a mean of 3,257 during 1910–1939.
Common Guillemot	Very large numbers taken but no accurate figures. Status of population poorly known, and change poorly understood.
Puffin	Important source of food for centuries; 150,000–200,000 still killed annually in the 1980s. Few early figures, but up to 50,000, although usually 20,000–30,000, taken in Westmann Islands and up to 10,000 taken on Grimsey in 20th century.

c) Faroe Islands

Fulmar	First colonised the islands between 1816 and 1839, and breeding on nearly all islands by 1900. Over 100,000 young were taken in the early 20th century, the harvest apparently starting in 1883. Psittacosis found in young in 1933, at which point catching Fulmar was banned.
Gannet	200 adults and 200 young taken in 1782, 300 adults and 600 young in 1862, up to 450 adults and 850 young taken in the 1930s; an average of 752 young taken 1936–1944.
Common Guillemot	Well over 60,000 taken per year, and probably more than 100,000. From mid-1940s shooting and snaring at sea much increased, and annual toll put at 100,000–150,000 in 1970s. Traditional fowling had little impact over the centuries, but shooting and snaring at sea, permitted throughout the year up to 1979, probably contributed to a population decline of perhaps 75% during second half of 20th century.
Puffin	Williamson (1948) suggested that in a good year the total catch 'must be in the region of 400,000–500,000'. The annual kill from fowling in the early 1900s was put at 270,000 by Salmonsen (1935, in Nettleship & Birkhead 1985), who noted that the harvest was 'now much reduced', although shooting and snaring at colonies had increased dramatically in the same period.

Table 8.1. *Numbers of four principal species of seabirds taken for food at three important seabird fowling sites, a) St Kilda, b) Iceland and c) the Faroe Islands. Sources Martin 1698, MacKenzie 1905, Gurney 1913, Fisher & Waterston 1941, Fisher & Vevers 1943, Williamson 1945 and 1948, Fisher 1952, Fisher & Lockley 1954, Nettleship & Birkhead 1985.*

Gurney (1913) also gave details of the harvest taken from other gannetries; up to 1885, the one at Bass Rock yielded 1,000–2,000 young birds annually, from an adult population of around 6,500 birds. The harvest ceased after 1885, but 1,800–2,000 eggs were taken annually into the early 20th century. The harvest on Ailsa Craig never exceeded 500 young birds, but eggs were collected from the more accessible sites right up to Gurney's day. Large numbers of other seabirds, chiefly adult Puffins and Common Guillemots, were also taken at this site. At the gannetries at Sula Sgeir and Sule Skerry Stack, between 2,000 and 3,000 and up to 1,200 young birds were taken respectively.

Fisher & Lockley's (1954) conclusions from their study of seabird fowling in these areas is worth repeating here. They concluded that

> *seabirds were a source of cheap and good food, which can be indefinitely enjoyed, provided a calculated harvest be taken from the cliffs which leaves a strong adult population behind. Further, this harvest can be large and yet not materially affect the size of the colony. Experience over some hundreds of years shows that large numbers of eggs can be taken from those species such as the Guillemot which readily lay replacements, and that from all others a crop of fat young can be taken most easily just before the young fly or swim away from the colony. Furthermore, it seems that it is safe to take up to half as many young as there are nests in the colony.*

Nevertheless, exploitation probably did influence these seabird populations. The Gannet provides a good example. Although the nine colonies around the British Isles in the 19th century could sustain the impact of the continual annual exploitation to which they were subjected, that exploitation prevented expansion. This is clear from the considerable increase in range and numbers of Gannets in the 20th century, when 12 new colonies were formed and numbers increased from *c.*48,075 pairs in the early 20th century (Gurney 1913, Fisher & Vevers 1941) to 187,800 by the mid-1980s (Lloyd *et al.* 1991), as exploitation progressively declined. Williamson (1948) made the interesting point that, despite the numbers taken, Puffins on Faroe could be readily induced to colonise new areas by digging artificial burrows for them, suggesting that exploitation was not impairing their status. Paradoxically, the steady decline in the harvesting of Puffins in British waters in the 20th century has coincided with a steep decline in their population (Nettleship & Birkhead 1985).

The northwest Atlantic

The exploitation of seabird colonies in northwest Europe differed sharply from events happening across the Atlantic, in what is now Canada and the northeastern United States. From the end of the 15th century, European fishermen, explorers and settlers found an astonishing wealth of fish, cetaceans and seabirds in this corner of the New World. They set about exploiting this wealth, without any regard for the future; it was regarded as an infinite and indestructible resource and treated accordingly, with the lack of any overriding authority to prevent such actions not improving matters. Seabird colonies came to be used as revictualling stations for voyagers, and voyages were planned with that objective in view (Nicholls 2009). The Great Auk *Pinguinus impennis*, being flightless, was the most vulnerable and easily taken species and was exploited for both meat and eggs. Later, they were also exploited commercially for feathers and oil. In 1785, Captain George Cartwright wrote in his journal of Funk Island, off the coast of Newfoundland (from Bent 1919):

> *Innumerable flocks of sea fowl breed upon it every summer, which are of great service to the poor inhabitants of Fogo* [an island off the coast of Newfoundland]*, who make voyages there to load with birds and eggs. When the water is smooth, they make their shallop fast to the shore, lay their gangboards from the gunwale of the boat to the rocks and then drive as many penguins* [Great Auks] *on board as she will hold ... it has been customary of late years for several crews of men to live there all summer for the sole purpose of killing birds for the sake of their feathers; the destruction which they have made is incredible.*

The surprising fact was not the extinction of this species, but that it took until the end of the 18th century.

The Great Auk was perhaps the most extreme example of rampant over-exploitation, but all seabirds suffered. The extremely numerous populations of other auks, particularly of Common Guillemots and Atlantic Puffins, in eastern Canada, the Gulf of St Lawrence, Newfoundland and the coast of Labrador, and the northeastern United States were ravaged by commercial egging and shooting in the breeding season, which extirpated many colonies in the 19th century.

In the early 20th century Bent (1919) surveyed much of the north coast of the Gulf of St Lawrence and of Labrador. He wrote of the Mingan Islands in the Gulf of St Lawrence 'where auks, murres, gannets and puffins formerly bred in great numbers, and which bear the name of the Parroquet Islands [presumably a reference to auks and their bills], are now almost devoid of bird life. The gannets have ceased to nest there and the puffins are almost wiped out'. Bent's reports show that this comment was applicable to the whole of the coast up to the Belle Isle Strait and along the northeast coast of Labrador to Cape Mugford, where he stated that he was forced to conclude 'that the large breeding colonies of the Alcidae had been nearly, if not quite, annihilated on the Labrador'. But large-scale shooting and killing of the much-reduced populations of Puffins continued into the 20th century (Nettleship & Birkhead 1985). Similarly, in Newfoundland, Bent found no Common Guillemot colonies on the north and west coasts, and few Razorbills *Alca torda*. He also noted that Audubon recorded a sharp decline in the Puffin population breeding in the Bay of Fundy by 1833; Bent found that they had virtually ceased to breed there.

The Russian Arctic

In the Russian Arctic, seabird colonies, known as 'bird bazaars', were of considerable economic importance, both as a source of fresh meat, eggs and feathers for local populations and as a source of export trade to other parts of Russia. This trade was principally developed on Novaya Zemlya, notable for its dense populations of Brünnich's Guillemots *Uria lomvia*, estimated at *c*.1.9 million birds in 47 colonies, and for its 30,000 Black-legged Kittiwakes in the 1940s (Uspenski 1958).

Uspenski (1958) had no doubt that coastal sailors and explorers in the Arctic had exploited these seabird colonies from the 15th and 16th centuries. But the first definite record of their exploitation dates from the early 19th century, when traders were taking guillemots and salting them down in barrels for their own use and for sale in Archangel. They also traded the birds' skins, which were made into beautiful 'furs' in much the same way that the Greenlanders treated Eider skins (see p. 88). Regular exploitation of these colonies began in the mid-19th century; in addition to the birds, this involved eggs, which were often processed as salted yolks and whites.

Until the Soviet government exerted full control over the area in the 1920s, much of this trade lay in the hands of Norwegians, and resulted in damaging over-exploitation. This was stopped in 1922; commercial activity was prohibited for eight or nine years, allowing the colonies to recover. Commercial exploitation began again in the early 1930s under the control of state trading companies; in 1932, 27% of the monetary value of exports from Novaya Zemlya comprised the eggs and meat of Brünnich's Guillemots, with this rising to 35% in 1933. In the Second World War, an important part of the food supplies of Archangel and Murmansk provinces came from the bird bazaars. Data on the total numbers of eggs and birds taken were incomplete, but they suggested that no fewer than 3,000,000 eggs and 500,000 birds were exported between 1930 and 1950 (Uspenski 1958).

Based on extensive research, Uspenski (1958) discussed the methods that should be adopted to develop the industry further while avoiding over-exploitation. That no improvement in the management of the industry was realised is clear from the account of Nettleship & Birkhead (1985). They wrote that in 1933–1934, before commercial exploitation took off again, the largest colony of Brünnich's Guillemots, at Guba Bezymyannaya, was estimated to hold 1,640,000 birds. Under exploitation this had declined to 600,000 birds by 1942, and to 290,000 in 1948. The population recovered somewhat with the stopping of the harvest and the declaration of the colony as a sanctuary, a recovery that appears to have been short-lived. Events at this colony were representative of Novaya Zemlya as a whole, although the timing of the start of exploitation differed from region to region.

The catastrophic result of exploitation as it was implemented as a quick and massive reduction in bird numbers, followed by permanently reduced populations or local extinctions, a situation not helped by the use of northern Novaya Zemlya as a test site for nuclear weapons from 1950. But one perhaps detects the malign influence of Soviet methods of state management here, with its culture of quotas and norms.

The 19th century in northwest Europe

Traditional methods of exploiting seabird colonies in some parts of the eastern Atlantic started to break down in the early 19th century, with the development of what Hawker (1893) described as 'rock shooting' (diaries) as an entertainment for tourists. On Lundy, D'Urban & Mathew (1895) observed that the Gannets on the Gannet Rock 'were so constantly harried by the Channel pilots that they shifted their quarters to another station on Lundy itself, where unfortunately the cliffs are not sufficiently precipitous to prevent their nests being plundered by the egg stealer ... the eggs are systematically taken, finding a ready sale at 1/- each [around £5 today] to tourists and others'. The colony was extinct by 1903, the birds moving to Grassholm, off Pembrokeshire. Gurney (1913) recorded that much the same thing happened at Bass Rock after Gannets ceased being harvested there. He noted that

> There was no demand for them, thus they became no longer worth protecting ... their principal value was thought to be only as a mark for sportsmen ... This sort of treatment must have gone on for a good many years, for when I was at the Bass in 1876 the lessee still bitterly resented the numbers shot after the 1st August, many of them even while flying with fish for their young ones, adding with emphasis that he had seen the sea strewn with dead Gannets, which the shooters did not take the trouble to gather up after they had shot them.

In northeast England, Kittiwakes and auks were also shot for sport in the breeding season by passengers or tourists in steamers from London, Newcastle and Leith. The birds shot were left behind, and the young left to starve (Muirhead 1895, Nelson 1907, Evans 1911). Macgillivray (1852) recorded that he had seen 'a person station himself on the top of the Kittiwake cliff of the Isle of May and shoot incessantly for several hours, without so much as afterwards picking up a single individual of the many killed and maimed birds with which the smooth water was strewn beneath'. It wasn't solely tourists in steamers that were engaged in this wanton activity. Cordeaux (1864) noted that

> since the railway has been opened along the East Coast, through Bridlington and Filey, hundreds of persons are brought down by 'cheap excursion trains' who otherwise would never have visited Flamborough, and boats and guns are too readily obtained ... it is scarcely possible to take a stroll along the top of the cliffs, during any fine day in the summer months, without hearing the constant report of guns fired from boats below.

Nelson (1907) noted that auks were shot in such large numbers at Flamborough Head that the traditional trade of the 'climmers' gathering eggs from the cliffs had to be discontinued because of the decline in the bird population. Protection under the Bird Protection Acts from 1869 saw numbers recover (Fielden 1870), and egging was eventually resumed.

This pointless target practice was not confined to northeast England. It also occurred along the south coast, on the Isle of Wight (Hawker 1893), and in Kent (Ticehurst 1909). In Sussex, Graham (1890) recorded that the local fishermen shot the Common Guillemots breeding on Beachy Head from below the cliffs and made them into pies; Dutton (1867) recorded that, a few years before, it was not uncommon for a party of two or three guns to kill 30 or 40 birds there in an hour or two, but numbers had declined markedly since. Christy (1890) also noted a marked decline of Little Terns *Sterna albifrons* in the Harwich area in the late 19th century 'on account of the persecution to which they are subjected by the visitors from Walton, who kill a great many during the nesting season, notwithstanding the provision of the Wild Birds Protection Act'.

Public concern at the impact of this senseless behaviour on seabird populations led to the Seabird Protection Act of 1869. But the close season set by the Act, April 1st to August 1st, was, as noted above, ineffective. So the Act was superseded by a series of Bird Protection Acts in the 1880s and 1890s, which did provide effective protection. Seabird colonies started to recover as a result.

Unrestrained collecting of eggs and killing of breeding adults in the nesting season were the major factors in the decline of many seabird populations in the 19th century, just as they were for most of the other groups of birds considered in this book. But fowlers dependent on seabirds for food usually faced the problem that their quarry was only available in the breeding season. Conservation of resources then depended on agreeing and adhering to a consensus on the level of harvest that was sustainable, and on what should be harvested. The uses to which harvested birds were put was also significant in this context. For example, Gannets were taken by the Labrador fishermen to bait their cod fish hooks. It was cheap bait, and sustainable supply was not an issue. The gannetry on the Bird Rocks in the Gulf of St Lawrence, which was estimated to hold 100,000 pairs in the early 19th century, was visited regularly by fishermen for this purpose. By 1898, the gannetry was reduced to no more than 750 pairs (Fisher & Lockley 1954). Fisher & Lockley remarked that erosion restricting the nesting area was also a significant factor in this decline, but that 'erosion of the Gannets kept ahead of erosion of the Bird Rocks'. The colony held 1,500 nests or more in 1954, when protected.

Traditional seabird fowling was starting to die out when the 1869 Act was passed (Fisher & Lockley 1954, Baldwin 1974), reducing its significance. The pressure of economic necessity was declining as new industries, particularly modern fishing methods and improved agriculture, replaced seabird exploitation, or people moved to areas where a more satisfactory living could be obtained. But the 1869 Act and its successors were not passed to curb traditional seabird fowling, for which they made exemptions, but to curb destructive abuses, in which they were successful.

The trade in seabirds

Besides being harvested for use by local communities, seabirds were also significant articles of trade. Archaeological records in Britain indicate that birds such as shearwaters, Common Guillemots and Razorbills were traded inland as early as the mediaeval period (Yalden & Albarella 2009). Puffins were exported from Britain to France from at least the 16th century (Bourne 1999b). This trade rested on the fact that these fish-eating birds were counted as 'fish' by the Roman Catholic Church, and therefore were permitted to be eaten on fast days and during Lent. Ray (1678) made this point in connection with the Manx Shearwater

Puffinus puffinus, for which there was a significant export trade from the Isle of Man to English epicures at this time.

Gannetries were valuable properties. One of the best documented colonies in British waters was the Bass Rock. Gurney (1913) gave considerable detail of the economic value of this colony noting that, in 1535, the Commander of the Fort was reported to have annually collected 400 gold pieces as the produce of the Gannets. In the early 17th century the figure was put at £200 per annum, and Ray (1678) recorded that, in 1661, the rent was £130 per annum, falling to £75 in 1678. In the mid-18th century, the rent and expenses of exploiting the colony amounted to £65–3s–1d and the produce realised £120–13s–5d, of which £108 was for 1,296 birds and £7–13s–5d for oil and feathers. The young Gannets were collected on four days a week in the season to ensure a constant supply to the Edinburgh market. One unexpected by-product of the Bass gannetry was the fish the birds brought in to their nests, which were collected in the 16th century by the men of the Garrison for their own use and to sell (Gurney 1913).

Feathers

An important part of the trade in seabirds was by-products, such as feathers and oil. Feathers were valuable. Hundreds of pounds of Gannet feathers were sold from the Bass annually. The feathers were worth 10 shillings per stone of 24 pounds (10.9kg); both Gurney (1913) and Mackenzie (1905) noted that a stone of feathers required around 80 adult birds to produce. Adult feathers were used for cheap pillows, cushions and feather beds. The feathers of young Gannets were used for beds; at least 240 birds were needed to make one bed (Baxter & Rintoul 1953). As late as 1876, the tenant of the Bass said it was worth his while to keep five or six women plucking Gannets at 1/6d (£7.50 today) per day throughout the season. Demand dried up at the end of the 19th century, and Gannets ceased to be harvested.

Feathers were a main staple at St Kilda. Gurney (1913) quoted Dr J. Wiglesworth, writing in 1902, who said that

> *Up to about 20 years ago enormous quantities of feathers (from Puffins, Gannets, etc) were exported as payment for rent, or in exchange for meal and other necessaries, and immense numbers of birds were killed for the sake of their feathers alone … the feathers indeed were at that time their chief source of livelihood.*

The feathers of Fulmars and Puffins may have predominated, but those of Gannets may have been more valuable, one Gannet yielding as much feather mass as 10 auks. In the latter half of the 19th century Puffins were mainly taken for their feathers rather than for food (Table 8.1), the plucked carcases being used as manure on the fields. Although the value of this trade declined in the early 20th century, there was steady demand from 1877–1897 and in 1894 the Factor for St Kilda was receiving around 200 stone of 24 pounds (more than two metric tonnes) annually, paying five shillings per stone for grey feathers and six shillings per stone for black ones (£23.75 and £28.50 today). These were worth seven and eight shillings (£33.25 and £38.00 today) respectively in Glasgow and Edinburgh (Gurney 1913). Gurney estimated that total weight as the equivalent of 90,000 puffin-sized birds.

MacKenzie (1949) noted that the raids on the puffinry on the Shiant Islands by the men of Lewis were mainly for feathers, and Lockley (1953) noted that the feathers and down from Puffins formed a significant article of trade and barter in Iceland and the Faroes. Williamson (1945) found that the soft white breast feathers of the auks were used extensively for stuffing pillows and quilts on the Faroes, and were also exported to Denmark before the Second World War, but the Faroese did not use the feathers of Fulmars, which had an objectionable smell. On St Kilda the feathers of Fulmars were mixed with those of other birds and sold to the Government for stuffing soldiers' pillows, being thoroughly fumigated before use, 'but in about three years the smell returns to them so strongly that Tommy Atkins refuses to rest his sleeping head on them until they have been again roasted' (Kearton 1909).

Little is recorded about the actual value of the trade in feathers in the western Atlantic, but it was clearly considerable. Besides the reference to the fate of the Great Auk on Funk Island (p. 136), J. A. Allen in Bent (1919) gave details, given to him by a frequent visitor to Funk Island in the early years of the 19th century, who said that

> these birds were formerly very numerous on the Funk Islands, and 45 to 50 years ago were hunted for their feathers, soon after which time they were wholly exterminated. As the auks could not fly, the fisherman would surround them in small boats and drive them ashore into pounds previously constructed of stones. The birds were then easily killed, and their feathers removed by immersing the birds in scalding water, which was ready at hand in large kettles set for this purpose. The bodies were used as fuel for boiling the water. (Bent 1919)

Bent noted that the remains of the huts and pounds were still present on the island when he visited.

It was not just Great Auks that were slaughtered; all seabirds on Funk Island were exploited for their feathers while breeding, except Arctic Tern *Sterna paradisaea*, which was regarded as useless for feathers or food, and Puffins, whose nests were generally impossible to dig out.

Razorbills were also keenly sought, particularly for their breast feathers, which were noted as extremely fine, warm and elastic. In America, Audubon noted that thousands of Razorbills were killed in Labrador for these feathers alone, with their bodies left strewn along the shore (Bent 1919), while the Greenlanders killed large numbers of Razorbills for food in February and March and dressed the skins to make clothing (Bent 1919).

In the Russian Arctic, Uspenski (1958) had little to say about feathers as a significant part of the yield of seabirds on Novaya Zemlya, although they were inevitably taken. Nevertheless, feathers were exploited from the beginning and, as noted on p. 137, skins were also taken and dressed to make fine 'furs'. This was a laborious process, and perhaps came to be regarded as uneconomic.

Oil

The oil or grease extracted from the Gannet was greatly esteemed in the past for its medicinal value. In the late 15th century it was regarded as a cure for gout and catarrh (a somewhat unlikely juxtaposition of complaints), and held to possess extraordinary healing qualities (Baxter & Rintoul 1953). Between 1764 and 1767, 10 Scots gallons of oil were extracted from the young Gannets on the Bass and exported, selling for £2–13s–5d. In the 19th century, when no longer used as a human medicine, it was sold at 2/6d (£12.50 today) per gallon as a lubricant, and a medicine for sheep before chemical sheep dips became available

(Gurney 1913). On St Kilda, Fulmar oil was used domestically in the islanders' lamps and medicinally, and Martin Martin, in the 17th century, remarked that it was the best preservative for iron against rust (Martin 1698). Gurney (1913) noted that in 1875 nearly 600 gallons (2,700 litres) of seabird oil was exported from St Kilda. But the trade declined steadily in the late 19th century as the price dropped from one shilling to 4^1/$_2$d per pint, and then lower still.

Great Auks were also important sources of oil in earlier centuries. Producing it meant heating the birds in large metal cauldrons. Lacking wood on nesting islands such as Funk Island, the oilmen of the 16th century onwards simply used the birds themselves, before they 'tossed their kin into the pot' (Nicholls 2009).

Methods of fowling

Much of what follows is based on Gray 1871, Mackenzie 1905, MacPherson 1897, Bent 1919, Williamson 1945, 1948, Lockley 1953, Fisher & Lockley 1954 and Norrevang 1986. Much of the skill involved in seabird fowling lay not in taking the birds but in actually getting to them, which often involved serious cliff and rock climbing. Many of the means of capturing seabirds were simple. For example young Gannets, known as *gugas* in Scotland, were clubbed and the bodies thrown into the sea below for waiting boats to collect. Martin (1698), writing of St Kilda in the late 17th century, noted that Gannets were also taken in horsehair snares. He recorded that the Gannets on Stac Lee were allowed to hatch their first clutches, but at other colonies on St Kilda the first clutches were taken to stagger the harvest of the young.

The young of auks were taken by hand just before they went to sea or, in the case of Puffins, dragged out of their burrows with an iron hook at the same stage. This also seems to have been the commonest method used to take young shearwaters. But on the Faroes, where it was safe to do so, young shearwaters were taken at night, dazzled with lanterns as they sat outside their burrows nerving themselves to go to sea (Williamson 1948). Alternatively, young auks were sometimes herded by boats into shallow water when they first left the cliffs in the Faeroes and killed by a well-directed stone attached to a fishing line. Simon (1952) quoted Dr Caius in the 16th century, who said that Puffins were also taken with ferrets.

Adult Puffins and Fulmars were taken in the Faroes and Iceland with the *fleygastong*, a net about three feet long (0.9m) and triangular in shape (much as a lacrosse net), mounted on a 12-foot (3.7-m) pole. It was used from traditional stances (an indentation on the cliff edge), known in the Faroes as *roks*. The fowler held the net inconspicuously beside him on the ground and swept it up and out to intercept Puffins and Fulmars flying along the cliff edge, a performance requiring skill and strength. The first birds caught were generally set up as decoys to lure others in. The *fleygastong* was not used elsewhere in the north Atlantic, except that it was recorded in Orkney around 1808, but it recurs in the north Pacific, in the Commander Islands, where it is used today to take Tufted Puffins *Fratercula cirrhata*. Elsewhere in the Scottish islands, flying Atlantic Puffins were knocked down into the water with a pole or the butt-end of a heavy fishing rod, to be collected in boats below. Some men became adept at this, taking upwards of 100 in a day, and Gray (1871) records that David

Bodan, the tacksman (bailiff) at Ailsa Craig, undertook for a wager to kill 80 dozen Puffins with a pole in a day. He won the bet.

These methods were not used on St Kilda, the place of the *fleygastong* being taken by a fowling rod. This comprised a long flexible rod tipped with a noose of horsehair plaited with Gannet quills, which was flicked over a sitting bird's head and pulled back toward the fowler, drawing the noose tight. This device was used to take auks, Gannets and Fulmars. A similar device was used elsewhere in western Scotland; Buckley (1892) recorded 1,500 Common Guillemots and Razorbills being taken on Handa by nooses at the end of poles in the 1880s.

For Puffins, the St Kildans also used a puffin-snare or gin, which consisted of a length of stout cord or rope, fastened to the ground at each end, with a series of horse hair nooses spaced along it. This was set where the Puffins had their daily assemblies and the Puffin's bump of curiosity soon led them to become ensnared, their struggles attracting others into the trap. On a good day a fowler with four or five of these traps might catch several hundred birds. This device was confined to St Kilda, but in the 20th century in the Faroes and Iceland floating rafts set with snares were sometimes used to trap auks at sea, the auks being attracted to them by stuffed decoys.

On St Kilda adult Gannets and auks were also stalked and taken by hand, and early in the season, before they laid, Guillemots were ambushed by fowlers hiding on their ledges overnight and catching them as they arrived in the early morning. They were attracted in the half light by the fowler holding up something white and then setting up those he caught as decoys. Up to 60 or 70 could be taken in this way, but it only worked once in a season at each site, the fowler working around a series of traditional stances.

Nets were also used in a more wholesale fashion. On Ailsa Craig, Gray (1871) noted that many Puffins were taken in nets spread over the breeding areas at night, the fowler returning in the early morning to take the catch; several hundred could be taken in this way. Puffins were also taken this way in Pembrokeshire. Greenlanders took Razorbills in winter in nets made of split whalebone.

In the Faroes, Guillemots were also taken with the *fleygastong*. When they first arrived fowlers rowed under the cliffs, made sufficient noise to startle and flush the birds and caught them as they dived out to sea. In early summer, non-breeding immatures were taken from boats as they flew to and fro at the foot of the cliffs. Later in the summer, when these birds moved higher up the cliffs, they were taken by fowlers lowered on ropes, who often spent several days on the cliffs doing this.

The gun

In the 20th century the gun came to be increasingly used for taking birds, and the shooting of adults at sea began to damage populations. Shooting and snaring in the breeding season in the 20th century has contributed to a decline of up to 75% of the Common Guillemot population of the Faroes, which was given full protection between April 1st and August 15th from 1980 as a result (Nettleship & Birkhead 1985). Uspenski (1958) recorded that small bore rifles were used in the auk colonies on Novaya Zemyla, which resulted in many birds being wounded and lost to both fowler and breeding population.

In Greenland, the shooting of Brünnich's Guillemot on the west coast since 1880 has resulted in drastic population declines, and the extirpation of many colonies. Since 1930 the population decline has been of the order of 35–50%, caused by the combined impact of summer and winter shooting and mass drownings in offshore salmon gill-nets. The shooting bag rose from 200,000 in the late 1940s to 300,000–400,000 in 1988–89, and salmon nets were accounting for a further 250,000–500,000 birds annually between the 1960s and 1976, when the fishery was closed. The birds shot were sold in local markets or sent to freezer processing plants for export to Denmark (Del Hoyo *et al.* 1996). But many were not used at all and were simply dumped (Hansen 2002).

One of the most serious recent assaults on alcids anywhere has been the winter hunt of guillemots in Newfoundland and Labrador. There, the traditional, low-volume subsistence hunting of the 1940s has evolved into a massive black-market and recreational sport-hunt, without bag limits. The birds shot are sold illegally, but residents of the province are permitted to shoot as many as they like between September 1st and March 31st. Annual kills have totalled up to one million birds, mainly Brünnich's Guillemots coming from the eastern Canadian Arctic, west Greenland, Spitsbergen and Iceland. This expansion of hunting has resulted from the economic growth of the province and the availability of high-powered speedboats and firearms, which have extended the reach and scale of such activities. Regulations to control this hunting were established in 1994–95 but have proved unenforceable in such a large and remote an area. It is almost certainly the cause of the 30% decrease of Brünnich's Guillemot in eastern Canada between the 1950s and early 1980s, and has doubtless contributed to the decline in west Greenland (Del Hoyo *et al.* 1996). Motor boats and modern firearms have also been important in the over-exploitation of seabirds in Greenland (Hansen 2002); Hansen found that the use of motor boats fitted with 85hp outboard engines meant that almost every ice-free area along Greenland's west coast could be reached from a town or settlement within three hours.

Seabirds as food

Virtually all seabirds are or were exploited as food somewhere in the world. In the North Atlantic and Arctic the principal species taken were the Gannet, Fulmar, Common and Brünnich's Guillemots, Razorbill and Puffin. The Great Auk was of major importance until the 17th century, and gulls, particularly Black-headed Gulls, were also exploited. Fulmars did not figure in the diets of many of the communities exploiting seabirds for food until the species began its great range-expansion in the early 19th century. It started to colonise the Faroes between 1816 and 1839, and spread to Britain outside St Kilda (where it had long bred) in 1878, and to Ireland in the early 20th century (Fisher 1952). Psittacosis appeared in the Faroe Islands population in 1936, after which the Government forbade use of the young, which were the source, as food (Williamson 1948). The same problem arose in Iceland not long after.

Fielden (1872) recorded that Kittiwakes were much esteemed for the table in the Faroes, where the inhabitants were 'constantly shooting them; we ate them and found them tolerable'. In the 1940s, however, the fledglings being of minor food value, were taken only when bad weather precluded more important work, though they were taken in larger numbers in seasons of scarcity (Williamson 1945). Kittiwakes eggs also provided a food crop in early summer. Williamson noted shooting only at one site, except for a few shot at sea by fishermen to use as bait. He also noted that the young were said to taste good and were usually eaten fried for breakfast. The species was taken in large numbers in Iceland (Fisher & Lockley 1954), but it was apparently not systematically taken for food in Britain. Nevertheless, Pennant (1776) observed that 'the young of these birds are a favourite dish in north Britain, being served up roasted, a little before dinner, in order to provoke the appetite; but, from their rank taste and smell, seem much more likely to produce a contrary effect'. Heligolanders shot large numbers of Kittiwakes in November and December, which they ate in pies (Gatke 1895).

Shearwaters and cormorants

Manx Shearwaters were taken just before they could fly at most of their north Atlantic breeding islands but, although much relished on St Kilda (Mackenzie 1905), they were not a major source of supply anywhere except on the Isle of Man, where they bred in huge numbers on the Calf of Man until the end of the 18th century (Fisher 1997). The islanders harvested them in early August and salted them down in barrels, later eating them boiled with potatoes (Pennant 1776). James Chaloner, writing in 1656, observed that 'the flesh of these birds is nothing pleasant fresh, because of their rank and fish-like taste; but, pickled and salted, they may be ranked with Anchoves, Caviare or the like' (in Ralfe 1905).

Bishop Wilson, in his *History of the Isle of Man* published in 1797, observed that 'great numbers, few years less than four thousand or five thousand, were taken'. He added that they were almost one lump of fat but 'they who will be at the expense of wine, spice and other ingredients to pickle them, make them very grateful to many palates, and send them abroad; but the greatest parts are consumed at home, coming at a very proper time for the husbandmen at harvest' (in Ralfe 1905). Despite the Bishop's figures for birds taken, Ralfe only gave one figure, for 1708, when 2,618 young were harvested and sold for £13–5s–6d. Ray (1678) noted that they sold for 9d a dozen, but that the trade was nevertheless worth £30 per year, suggesting that up to 10,000 were taken. But Fisher (1997) observed that the main profit from the birds came from oil and feathers, and it is not clear whether Ray's total included this. Pennant (1776) noted that Manx Shearwaters were much valued in the Orkneys, both for food, salted down for winter provisions, and for their feathers.

Despite the scale of exploitation on the Calf of Man, the colony continued to yield large numbers of birds into the 19th century. Its demise was due to a combination of over-exploitation and predation by rats, which arrived on the Calf of Man from a Russian ship wrecked in 1786 (Fisher 1997).

Cormorants and Shags were little exploited, although the Faroe Islanders regarded young Shags as the best of the sea-fowl they caught for the table, at least in the early 19th century when the Rev. G. Landt described them as tasting nearly as good as roast hare (Williamson 1948). This is an analogy that occurs surprisingly frequently in the 19th-century literature when discussing water birds as food, and it was repeated by Graham (1890), when describing shooting Cormorants, known as *scarts*, in Iona and Mull. He suggested that this was a widely held opinion in the Hebrides. Edward Topsell, in the 17th century, also noted that Shags were good eating, comparing them to wild geese (Harrison & Hoeniger (1972).

The Gannet

Young Gannets were highly regarded as articles of food in eastern Scotland. Gannets from the Bass Rock were served to the kings of Scotland in the 16th century, and Sir Robert Sibbald, in the late 17th century, noted that 'those who pamper their appetites most, cannot by their skill produce any flavour so delicate of meat and fish mingled as this bird has when roasted. Hence among our countrymen the full grown chickens [juveniles] are held to be delicacies and fetch a high price'. Plucked and prepared for the market by being flayed and gutted, they fetched 1/8d each (Rintoul & Baxter 1935), although prices declined to a shilling or less in the 19th century, a decline reflected in the rental value of the colony, which had fallen to £30–35 per year in the 19th century (Gurney 1913).

Gannets from the same source were sold in Edinburgh in the 17th and 18th centuries, and were sent to England from there and from Ailsa Craig. In England they were served in many of the less salubrious eating houses in cities such as London, Sheffield, Newcastle, Birmingham and Manchester, a trade that continued until the late 19th century (Gurney 1913, Baxter & Rintoul 1953). Gurney remarked that whether the consumers knew what they were eating was doubtful, the birds generally being known as geese. He also noted that Ailsa Gannets were never esteemed as of the same value as those at the Bass Rock and, locally, were only sold for consumption among the lower classes, together with Puffins and Guillemots, with these birds being eaten by those who could not afford mutton and beef. Baxter & Rintoul (1953) noted that Gannets 'were hawked about the countryside in carts, fresh as well as salted, or retailed at various Lothian markets to anyone who would buy them'. They were relished by the immigrant Irish labourers who came to East Lothian in the 19th century to gather the cereal harvest. During August a brisk trade went on at Canty Bay, 'where 200 young Gannets, landed in boats from the Bass and afterwards cooked in brick ovens, are said to have been sometimes consumed at a single feast' (Gurney 1913).

There are interesting divergences of opinion about the palatability of seabirds as food. Mackenzie (1905) made the interesting point that all the seabirds visiting St Kilda were only good eating when fat. Williamson (1948) had a far more enjoyable encounter with Gannet-flesh than the historical record would suggest. In the Faroes, he wrote 'There is much meat on a single bird, and ... I thoroughly enjoyed it, finding the meat very tender and quite devoid of the objectionable 'fishy' flavour I had expected'. He remarked that the people of Mykines held a Gannet to be worth as much as a goose and, in fact, the Gannet replaced the domestic goose on the tables of that island. Nor did he note that any special preparation was used. Lockley (1953) agreed with Williamson, recording that the flesh of young seabirds was usually delicious eating, with that of the young Gannet like tender beefsteak, whilst the slight fishy smell disappeared in the cooking; roasting or frying seemed best.

Similar differences of opinion arise with the Puffin. Pennant (1776) remarked that 'their flesh is excessive rank, as they feed on sea weeds and fish, especially sprats: but when pickled and preserved with spices, are admired by those who love high eating'. Richard Carew in his survey of Cornwall, written in 1602, noted that they were 'kept salted, and reputed for fish as coming nearest thereto in their taste' (Lockley 1953). Puffins from the Isles of Scilly were noted in 1673 as having an ill taste, and being a food fit only for servants (Newton 1896). But this opinion was not universal; on the Isle of Man, Puffins were regarded as excellent food in the late 18th century, good and nourishing (Ralfe 1905). Mackenzie (1949) also recorded that the Lewis men who raided the puffinry of the Shiant Islands in the late 19th century enjoyed big pots of boiled Puffins for their dinners as a welcome change from the usual fish diet, although the birds were taken mainly for their feathers. Williamson's (1948) opinion was much more enthusiastic, for he found it a delicious meat, rich and tender and very tasty. This opinion was endorsed by Fisher & Lockley (1954) and by Lockley (1953).

Why such differences of opinion should arise is unclear. The birds were usually salted down or smoked, and Williamson gives no hint of special methods of preparation. Lockley said that Puffins were eaten either boiled or braised and made excellent soup. It is possible that early authors made assumptions, having declined to risk it (as Williamson did the Shag!).

Williamson (1948) also noted that Common Guillemots and Razorbills were often eaten in the same manner as Puffins in the Faroe Islands, sometimes stuffed with a pudding containing currants and raisins. Although meatier, Guillemots were inferior eating to both Puffins and Razorbills. Fulmars he found tasteless, but they were eaten in the Faroes as a source of fresh meat in winter, the time that they were mainly taken. But on St Kilda they replaced the Puffin as the most important sea-fowl for food in the 19th century, the Puffin being mainly harvested for feathers at that time. Fisher (1952) devoted an appendix to the widely varying opinions on the Fulmar as food, ranging from the Greenlanders at Disko, who found them too foul to eat, to the St Kildans who enjoyed young Fulmar-flesh. Fisher's account gave some indication that the palatability of the bird varied with its diet.

The Black-headed Gull

Black-headed Gulls were a special case among seabirds in this context. They often bred on easily accessible inland marshes and pools, or similar island sites along low-lying coasts. Stevenson (1871–72) aptly described them as a land bird rather than a seabird in the breeding season. In England such colonies were valuable properties in the past, carefully managed for their yield of young and eggs.

In the archaeological record for Britain, Yalden & Albarella (2009) noted the species in 18 sites, mostly from the mediaeval period, and remarked that it was apparently much less numerous then than it is today. Nevertheless, in England and Wales, a number of substantial Black-headed Gull colonies are recorded in the old literature (these are listed in Appendix 4). That list is also incomplete. Christy (1890), for example, drew attention to at least six sites on the Essex coast with names such as Cob or Pewit Island (see below); this indicates that they were once occupied by Black-headed Gull colonies, long since vanished. Stevenson (1871–72) and Stevenson & Southwell (1890) refer to a number of long-abandoned sites in Norfolk lost to drainage and agricultural improvement in the 18th century. Sites in the Fens and in lowland marshes in Yorkshire, recorded by Pennant (1776) and by Chislett (1953), were similarly lost. Gurney (1921) suggested that the species nested in Cornwall in the 17th century but it bred only sporadically in the county from the mid-19th century (Penhallurick 1969).

Although the information about the exploitation of the colonies is somewhat fragmentary, such exploitation was certainly extensive. The young comprised the main crop up to the end of the 17th century but went out of fashion in the 18th, when the colonies were increasingly cropped for eggs.

Young Black-headed Gulls, known as puetts, puits or pewits, were esteemed as table birds in the mediaeval and early modern periods, often appearing in household accounts, as gifts between gentry establishments, for example. Willughby & Ray's (1678) account of harvesting the young at the famous gullery at Norbury in Staffordshire probably describes the typical method of taking them. The colony was on a large island in a marshy pool, and their account records that

> *When the Young are almost come to their full growth, those entrusted by the Lord of the soil drive them off the Island through the Pool into Nets set on the banks to take them. When they have taken them they feed them with the entrails of beasts, and when they are fat sell them for four pence or five pence apiece. They yearly take about a thousand two hundred young ones.*

Gurney (1921) showed that the fowlers used pens made of wattle hurdles to hold the young during the drives (Figure 8.1). Ray underestimated the total crop, since McAldowie (1893) quoted figures from Dr Robert Plot, writing in 1686, showing that 2,200–3000 were taken annually, making a profit of up to £60 per year. Plot noted that 50 dozens were sometimes taken at one drive; up to three drives were sometimes taken in a morning, and three days around the beginning of June were appointed for the harvest. He also recorded that the habitat of the nesting islands was managed every winter by controlling the vegetation. The additional cost of constructing houses for holding and purchasing food for the gulls appears in various household accounts of the 16th and 17th centuries.

As Ray's account shows, the young commanded high prices. Lennard (1905) noted that 6d apiece was the going rate in Sussex, while Gurney (1921) recorded that some Oxford colleges in the late 16th century paid up to between 2/3d and 3/4d each. They were listed at 10d in the London market in 1633 (Jones 1965).

In the 19th century, the collection of Black-headed Gulls' eggs became increasingly excessive, resulting in the loss of colonies and a marked population decline. One reason for this was the rising demand for Lapwing eggs, which had contributed to a marked decline in that species in the late 19th century (Shrubb 2007). Black-headed Gull eggs are very similar in size and appearance, and were widely sold as Lapwing eggs, as Hope (1990) attested, but they were markedly inferior in flavour; Yarrell (1845) remarked that they were somewhat

Figure 8.1. *Harvesting young Black-headed Gulls, or puets, at Norbury, Staffordshire, in the 17th century. From a 17th-century print in Gurney (1921).*

like duck eggs. That this was a profitable trade was shown by Cott (1953), who cited the example of a Cambridge poulterer who was handling up to 1,000 Black-headed Gulls' eggs weekly in 1949. The poulterer sold eggs early in the season at 8d each (around £1.60 today); at the height of the harvest the eggs went for 6d, declining to 4d (£1.00 to 67p today). James Fisher reported that one London club was offering its members Black-headed Gulls' eggs at 9d to 1/- each (around £1.50 to £2.00 today) in 1951.

The collection of eggs was restricted with the passing of the Birds Protection Acts in the 1880s, and the species started to increase and expand its range from the end of the 19th century, having reached a nadir in the 1880s, when there were fears for its survival as a British breeding species. But such protection was not always effective. In Lancashire, Oakes (1953) noted that the increase had not been maintained due to the systematic plundering of eggs at all unprotected sites as a result of food shortages in the First and Second World Wars. The 1954 Protection of Birds Act made the collection and sale of eggs for food legal again, and Cohen (1963) observed that eggs were then gathered on a commercial scale in the large Black-headed Gull colonies in Hampshire.

In Scotland, Black-headed Gulls were always rather more numerous and widespread than in England and Wales, but there appear to be few references to the taking of young for food. The only one that I have found is in Sim (1903), who refers to a report in 1641 that soldiers in Aberdeen took eggs and young. But eggs were widely collected which, as in England, contributed to a marked decline in the 19th century. For example, the ancient colony on the Loch of Strathbeg was deserted due to excessive egging (Baxter & Rintoul 1953), and the series of *Vertebrate Faunas of Scotland* edited by Harvie-Brown contain numerous other examples of colonies heavily exploited for eggs.

The Black-headed Gull's history in northwest Europe is similar. Cott (1953) showed that it was one of only nine species for which the collection of eggs fell into his category of 100,000–1,000,000 per year in the 20th century. Hagemeijer & Blair (1997) noted an increase in the late 19th century associated with a decline in egging. But eggs continued to be heavily exploited well into the 20th century in some areas. In the 1940s they were regularly gathered in the Netherlands, and used especially for cakes. Eggs were collected into the 1950s in Denmark, both for local use and for export; young birds were also taken in the past there (Cott 1953).

Other seabird eggs

The exploitation of seabirds' eggs was a worldwide industry and it is only possible to summarise it briefly here, with particular reference to the parts of the world and the species already dealt with in this chapter. Its scale was even greater than the fowling already described, and unbridled egging was probably more destructive of the colonies of some species. A detailed analysis of the exploitation of wild birds' eggs worldwide was published by Cott (1953, 1954) and I have drawn heavily on his account.

Cott listed nine bird species that he estimated yielded an annual crop of 100,000–1,000,000 eggs worldwide although, as he stated, accurate figures throughout any species range could not be known; one species, the Sooty Tern *Sterna fuscata*, yielded 1,000,000–10,000,000. In addition the Common Eider and the Lapwing yielded 100,000–1,000,000 annually in the 19th century. Of these species all but one were seabirds. In the North Atlantic and Arctic the species concerned were Black-headed Gull, Herring Gull *Larus argentatus*,

Arctic Tern and Common and Brünnich's Guillemot. Common Gull *Larus canus,* Lesser Black-backed Gull *Larus fuscus* and Kittiwake also yielded significant crops of eggs as, before the 20th century, did the Razorbill.

Terneries and gulleries were widely exploited for eggs on both sides of the Atlantic, often without any regard for the conservation of the colonies. Terneries were particularly vulnerable, so often being in easily accessible sites. Many were severely reduced or exterminated in North America by a combination of egging and shooting for the plumage trade (Bent 1921). In Britain, abundant tern colonies in East Anglia received similar treatment, with indiscriminate egging and, as soon as the close season ended in early August, shooting by 'sportsmen' from neighbouring towns. This reduced the colonies of Common *Sterna hirundo* and Little Tern *Sterna albifrons* along Norfolk's north coast from hundreds of pairs to tens or fewer by the end of the 19th century (Stevenson & Southwell 1890). The abundant colonies of Sandwich Terns *Sterna sandvicensis,* first described by Latham in 1784, had been extirpated by egging in the area of its type locality in Kent by the mid-19th century (Ticehurst 1909). Latham also described the Common Tern as more numerous in Sussex and Kent than anywhere else in Britain, but colonies of this bird were largely destroyed in the 19th century by unregulated egging (Knox 1849, Ticehurst 1909, Cott 1953). Walpole-Bond (1938) described it as one of the rarest breeding birds in Sussex.

Plundering auk eggs

Before the 20th century Cott also placed the Common Guillemot in his category of 1,000,000–10,000,000 eggs annually, largely as the result of the scale of egging on the coast of Canada and the northeast United States. The principal cause of the destruction of auk colonies there was the commercial egging industry (Bent 1919). This was organised mainly from Nova Scotia, Maine and Newfoundland. Auk eggs, particularly of Common Guillemots, were systematically collected from all the colonies, from Sept-Îles on the north shore of the Gulf of St Lawrence up to Nain on the Labrador coast. To ensure fresh eggs, crews landed and destroyed all the eggs they could find and then returned later to collect the fresh crop (Nicholls 2009). Because of this staggeringly wasteful method, most of the Common Guillemot and Razorbill eggs gathered for sale were from second layings, and eggers even returned to take third clutches as well. Thus the birds had few opportunities to rear young. This had a particularly serious impact on the Great Auk, which did not relay after losing its single egg (Serjeantson 2001) – a major factor in the bird's extinction.

The scale of the egg-harvest is indicated by some figures for the north shore of the Gulf of St Lawrence, where Common Guillemots were still sufficiently numerous in the late 1800s to permit an estimated harvest of 750,000 eggs per year. They were sold in the northeast coastal states of America at between one and three cents each. The trade died out at the end of the century, when too few birds remained for it to be profitable. With the passing of the Migratory Bird Treaty of 1916 between Canada and the United States, federal bird sanctuaries were established to safeguard the surviving small populations. Such protection has allowed a slow recovery in the populations of both Common Guillemots and Atlantic Puffins, although in neither case to anything approaching the myriads that once existed (Cott 1954, Nettleship & Birkhead 1985).

Right around the Arctic, the principal egg species was Brünnich's Guillemot. Large numbers of its eggs were harvested, for example, in Greenland and Spitsbergen (Cott 1954).

and Uspenski (1958) suggested that no fewer than 3,000,000 eggs of this species were taken on Novaya Zemlya between 1930 and 1950; in 1933 342,500 eggs were taken from the Bezymyannaye colony alone. Uspenski discussed in some detail the methods that should be followed to develop this harvest on a sustainable basis. Problems arose particularly with the timing of egg-collection. Many of the first-eggs taken were partially incubated, and either discarded or used as dog food; commercial collection for human consumption rested heavily on second layings. But Uspenski found that less than 2% of Brünnich's Guillemots laid a third egg, so this pattern bore heavily on the birds' productivity; the results of such exploitation were, as Nettleship & Birkhead (1985) noted, catastrophic.

In northwest Europe, eggs were widely exploited at all the main seabird stations, particularly those of the Common Guillemot. For example, Williamson (1945) recorded that in the Faroe Islands, up to 80,000 Guillemot eggs were taken per week on Skuoy, and Hestur and the Dimunar also provided large numbers. Most were preserved in a mixture of water, salt and peat ash for winter use, but many were also exported through Torshavn. Williamson remarked that he had tried them hard-boiled, but their appeal was mainly to the eye. He did not note the scale of the total harvest, but Cott (1954) suggested, on the basis of Williamson's figures and assuming a six-week collecting season, that it could have been as many as 500,000 eggs per year.

In Iceland, too, considerable numbers of Common Guillemot eggs were harvested but, by 1930, the colonies on Grimsey had been so reduced, apparently by epidemic disease (Congreve & Freme 1930), that egging-boats stopped calling. But guillemots' eggs are still substantially collected in Iceland today, and are on sale in supermarkets and other food shops (R. Lovegrove, pers. comm). Icelanders started also collecting Fulmar eggs when the harvesting of young was terminated due to psittacosis. Some Fulmar eggs were also taken on St Kilda, but generally the eggs of this species were not cropped, despite their being regarded as the most palatable of all seabird eggs (Fisher 1952), because Fulmars do not lay again after losing their egg.

On St Kilda, eggs, particularly those of the Common Guillemot, were an important crop. During about three weeks in 1697, Martin Martin's party were provided with 18 Guillemot eggs per man per day, in addition to eggs of other species (Martin 1698). During the period 1829–43 Mackenzie (1905) saw, on two different days, 17 and 14 baskets of eggs landed from Stac Biorrach, each containing some 400 eggs. In 1902, a favourable season, the harvest amounted to around 6,000 eggs. The birds started laying about May 15th or 16th, and collecting started on the 18th, with second clutches being collected 18 days later (Wiglesworth 1903, in Cott 1954).

'Climmers'

At seabird cliffs in eastern and southern Britain, where there was no regular exploitation of the birds for meat or feathers, eggs were harvested, most famously along the cliffs of Bempton, Buckton and Speeton in Yorkshire. Again, the species most affected was the Common Guillemot, but the eggs of Razorbills and Kittiwakes were also taken. Generally, however, the eggs of the Razorbill, although just as palatable as the others, were a less important crop because their nests were often inaccessible, while Kittiwake eggs were too fragile to transport any distance. Nelson (1907) gave a detailed description of the activities of the men (know locally as 'climmers') who climbed for the eggs on these cliffs (Figure 8.2); Nelson himself

151

Figure 8.2. *'Climmers' collecting auk eggs on Bempton cliffs in Yorkshire. From Nelson (1907).*

had often taken part. The right of gathering them belonged to the farming tenants of the adjacent land, who in turn conceded the privilege of collection to the men who worked for them, although the cliffs at or near Flamborough were worked by the local fisherman. In the 19th century between two and four teams worked the cliffs, dividing their length into sections worked exclusively by each. About 30 descents per day were made; the work was arranged so that each section was cleared bi-weekly, each portion being cleared every third day to ensure a constant supply of fresh eggs. In the 1860s and 1870s climbing started on May 12th, but by the end of the century it started 7–10 days later. Climbing ended in the first week of July, but in 1904 this date was moved so the season ended on July 1st.

The average daily take by each team was 300–400 eggs, which suggests a total crop of 70,000–80,000 eggs. In 1884, however, Cordeaux (1885) recorded that one of the four teams working took 30,000 eggs, mainly Guillemots', in 52 days. As local fishermen were also egging, the total crop that year may have amounted to 130,000 eggs, of all species. Cott (1954) noted that the total crop had declined to 18,000 in the 1930s, and 10,000 in 1949. First clutches were the most productive, but second and third clutches were also taken. When any portion of cliff became unprofitable it was left for two years or more until numbers had recovered. Nelson (1907) also remarked that there were many dangerous parts of the cliffs that were never climbed, and there the birds could rear their young undisturbed, ensuring a supply of fresh breeders. Eggs were sold at 6d per score in 1834, and at the end of the century for three farthings to a penny each. Bempton eggs were renowned for their variety and beauty, and well-marked or unusual eggs were set aside for collectors who, at the end of the century, would pay between five and ten shillings (up to £45.50 today) for good specimens.

Cordeaux (1864) suggested that egging was seriously reducing the avian population of these cliffs, although his account actually implied that the decline was largely the result of shooting by tourists (see p. 138). The population certainly recovered with the passing of bird protection legislation from 1869 (Fielden 1870). Nelson (1907) considered that little diminution of the population had occurred in his time. It also has to be added that, by Nelson's and Cott's accounts, egging had persisted at this site for at least 250 years without leading to the extermination of the colonies.

Similar egging traditions were followed at St Abb's Head in Berwickshire, where Muirhead (1895) observed that vast quantities of Common Guillemots' eggs were formerly taken by fishermen, and sold at 4d a dozen in 1837 and a penny each around 1846. Towards the end of the 19th century Guillemot numbers had declined considerably. 'Climmers' also operated on the Sussex cliffs, where Walpole-Bond (1938) noted that 'towards the middle of the 19th century the Guillemot was sufficiently numerous at Beachy Head as to excite the cupidity of certain locals – adept 'climmers' of course. And these in the course of a single morning often took twelve or fourteen dozen eggs, which found a ready market at six pence apiece'. The species largely disappeared from Sussex as a breeding bird as a result of cliff-falls destroying its nesting ledges later in the century, as it did from Kent.

Auks' eggs were also taken on the Farne Islands, on Steep Holm in the Bristol Channel (Lewis 1936) and on Lundy, but although Guillemots bred on the chalk cliffs of Kent up to the mid-19th century, Ticehurst (1909) could find no evidence that eggs had ever been systematically collected there. Nor does there appear to have been any tradition of egging auk colonies along the English coast to the west of Sussex.

Industrial egging

Seabird eggs were not just collected for food; large numbers were used for a variety of industrial purposes. Thus Uspenski (1958) noted that egg yolks, preserved in potassium dichromate, were widely used in the tanning industry, whilst the yolks and whites of spoiled eggs were a valuable raw material in soap-making, and powdered eggshells were an excellent source of minerals for farm animals. Many of the eggs collected on the Bempton cliffs were sent to Leeds, where the albumen was used in the manufacture of patent leather (Nelson 1907). In the southwest, eggs were used in the refining process at sugar refineries in Bristol, and they were used in a number of areas for clarifying wine. And, in Pembrokeshire, farmers in the St Davids area fattened their calves in the summer on a custard made from Common Guillemot eggs taken on Ramsey (D'Urban & Mathew 1895).

Egging remains a problem in some countries. Hansen (2002), for example, described it as 'the overwhelmingly dominant reason' for Greenland losing the world's largest colony of Arctic Terns, at Gronne Ejland, Disko Bay, which once comprised up to 80,000 pairs. He quoted Salmonsen (1967), who said that around 100,000 eggs were taken annually before the 1960s, an impossibility by 1967. Egging, combined with constant disturbance at the breeding grounds, led to a steep decline, and a census in 1996 found only 5,000 pairs remaining. During the census, up to 200 people were recorded visiting the colonies to take eggs. By 2000, a partial census there found no terns' eggs or chicks at all.

Hansen also noted a decline of 37% in just seven years in the population of Brünnich's Guillemots at Greenland's most southerly colony at Ydre Kitsissut due to egging, at what is supposedly a protected site. Hansen stressed that the constant disturbance of colonies was as important a factor in reducing populations as collecting eggs. Conservation laws have proved quite ineffective in this part of the world.

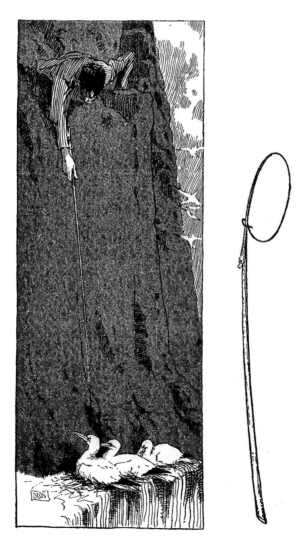

Figure 8.3. *Snaring Gannets on St Kilda, and the fowling rod used. From MacPherson (1897).*

CHAPTER 9
Passerines

Throughout history, passerines have been eaten by all classes of society. Even the smallest species were taken; MacPherson (1897) found bunches of Goldcrests *Regulus regulus* and Wrens *Troglodytes troglodytes* for sale in the Italian markets. One of the marked features of MacPherson's accounts is the extraordinary amount of ingenuity and effort expended to catch such small birds as tits and warblers. The latter were particularly esteemed for the table around the Mediterranean. Indeed, in England, in the period after the Black Death in the mid-14th century, the consumption of small birds increased despite their limited food value; eating them was held to confer the benefit of good health (Serjeantson 2006).

As noted in Chapter 3, there were no laws in England that prohibited the trapping of songbirds so long as they were not taken on the Lord's demesne, in his warren or in his park (Thirsk 2007). How much the Game Laws, promulgated from 1671 onwards (see p. 42), interfered with the countryman's ability to trap such birds is unclear. In England, however, I have little doubt that parliamentary enclosure, particularly of the commons, from the mid-18th century had a far greater impact, by denying the rural poor access to the important resources of grazing, fuel and other materials that the commons provided, and which were an essential part of their economy; these probably also included free sources of food. Thirsk (2007) observed of the period before 1760 that 'we never catch a glimpse of birds in the cooking pot of common folk, but we can take them for granted in the countryside'. Hope (1990) was of the same opinion. Both these authors also commented on the value of wild birds as a source of year-round fresh meat. The scale of rural distress that resulted from parliamentary enclosure in England is clear from reading Cobbett (1957). Even Arthur Young, the arch proponent of enclosure, came to the view in later life that much more thought should have been given to the impact on rural communities of the wholesale way in which the enclosure of the commons was implemented.

Hobusch (1980) remarked that, in Germany, hunting in the territories of many towns was regulated by their councils, and the snaring of songbirds was open to many townsfolk. Thus from the 17th to the 19th centuries the hunting of songbirds there became the hunt of the common man and 'with net, limed rod, snare, sling, ring net and decoy, birds were caught for centuries' (see Plate 6). Thousands of small birds were caught over decoys, particularly in autumn, plucked, put on skewers and sold in the markets. Small birds with dumplings was a Thuringian specialty; larks, Fieldfare *Turdus pilaris* and Quail *Coturnix coturnix* were in great demand.

Gatke (1895) discussed the Heligolanders' liking for thrushes, observing that

> *in Heligoland … almost everything finds its way into the soup pot, or rather the stewpan; not only every species of thrush, but also, by preference, Larks, a stray Wood Pigeon, Golden Plover, Peewit, Landrail, and the like. Hardly anything is roasted. For my part, I can only advise everybody who catches birds in sufficient numbers for once in a way not to roast his … Fieldfares, but, by way of a trial, to confide some forty or fifty, according to requirements, to the soup pot.*

Seed-eaters and pest control

The trapping of seed-eaters such as sparrows for food was a side-benefit of pest control. Hobusch (1980) observed that in 18th-century Germany, birds were often considered destructive vermin and hunted by order of the authorities. In Prussia, for example, as the result of orders made on December 12th 1721 and January 8th 1731, farmers, smallholders and crofters had to deliver so many sparrows' heads to the authorities, which led to 359,928 sparrows being destroyed in 1736; persecution on this scale continued until 1767. As Hobusch remarked, no one knows how many other species were destroyed in this campaign.

'Sparrow clubs' were widespread in Britain in the 19th century. In an attempt to control the impact of House Sparrows *Passer domesticus* on cereal crops, vast numbers of sparrows and their eggs were taken by such clubs (Lovegrove 2007), but there is no evidence that this had any impact whatsoever on a species universally described in the 19th century as a too-abundant pest. Bat-fowling (Figure 9.1) was a common activity. Birds were flushed from their roosts at night into a light and caught with a tall, folding hand-net as they fluttered, dazzled, round the light. Nets were not always used. In the 17th century the birds were knocked down and killed with heavy bushy poles (Harting 1871) – hence bat-fowling, the 'bat' referring to the implement used, rather than the quarry or the time of day. Many other passerines besides sparrows were caught by this method.

Ticehurst (1909) recorded the use of 'sparrow bottles' in parts of Kent and Sussex. These were artificial clay nest-sites shaped like a squat milk bottle, with the base left half open to allow the contents to be removed. They were hung beneath the eaves of houses to induce sparrows to nest there, and facilitate the collection of eggs and young, which qualified for bounty payments from the parish. Montagu (1833) noted the same arrangement in parts of London but recorded that their purpose was to *prevent* the birds from nesting under the eaves, where they dug out the mortar to make larger nest holes. Cott (1954) recorded that

Figure 9.1. *Bat-fowling, which had nothing to do with bats. From a 19th-century engraving in Yarrell & Newton (1882).*

early literature contained many references to the supposed value of sparrow flesh and eggs to promote sexual desire, suggesting another reason to encourage their proximity.

Farmers encouraged their workers to use riddle-traps to catch sparrows, finches and buntings to protect their corn ricks (stacks of sheaves), from which flocks of Corn Buntings *Emberiza calandra* could strip the thatch and let in water to spoil the crop (Knapp 1829). This provided men and boys with some amusement and some useful protein in the form of House Sparrow or Corn Bunting puddings (Shrubb 2003).

Macgillivray (1837–52) remarked that Corn Buntings became remarkably fat in winter, at which time they became superior as a food item to most small birds, including Skylark *Alauda arvensis*. Bewick (1826) also recorded that Corn Buntings were shot or netted in great numbers and probably sold as larks, while Smith (1930–38) noted that in the 19th century they were shot 'abundantly' in farmyards in north Staffordshire in winter.

Fowling methods

An extraordinary variety of hand-nets, bow-nets, drop-traps, spring-traps, leg-traps and so forth were devised for catching small birds, but nets (particularly clap-nets), bird lime (often in the form of lime sticks) and snares were the main methods of taking them; the gun is increasingly used today. Decoys and call-birds were important in attracting birds to the fowler, often combined with small owls to draw mobbing flocks within range (see Plate 6). Good call-birds were valuable; Daines Barrington (in Pennant 1776) recorded a price of £5–10s given for a good Linnet *Carduelis cannabina*, for example. There was a considerable trade in call-birds and decoys and a great fair of decoy birds was held annually outside Florence on September 28th (MacPherson 1897).

Baits were important tools in trapping birds. Hemp seed, for example, was scattered in the netting area to encourage seed-eaters to settle. Many passerines were caught in snares baited with berries, and large numbers of thrushes, particularly Fieldfares, were caught in snares baited with grapes in vineyards in autumn (Brusewitz 1969).

Elaborate permanent constructions were built for netting birds in Germany and, particularly, Italy. In Italy they were known as the *roccolo,* or the *breccianella* (which was simply a larger version). Both were usually rectangular enclosures, semi-circular at one end and open at the other, walled with netting 12m high. Trees were trained on trellis work around the outside, with some often left standing within the roccolo (Figure 9.2). The interior of the breccianella was sometimes planted with some low bushes to attract thrushes, with an area of ground made into an attractive feeding-site for seed-eaters. The open end was occupied by the fowler's house or watch-tower, and birds were attracted into the catching area by the usual combination of decoys, call-birds and bait. When birds settled they were flushed into the nets, in many cases by throwing a woven racquet over them and shouting. The Germans made similar permanent catching-sites but used clap-nets to take their birds. Many Italian estates also had permanent clap-netting sites or fowling fields, known as *largos,* where up to five pairs of nets could be operated by one man in a centrally placed hut. Hobusch (1980) noted that some 5,325 permanent netting stations were operating in Italy in the 1960s, when 3,232 bird trappers were also licensed.

In Russia, fowlers sometimes planted long strips of ground with berry-bearing shrubs, typically rowan, elder, bird cherry and juniper, to attract birds such as thrushes and Waxwings *Bombycilla garrulous.* Nets were set across and along the sides of an open space left in a clearing in the plantation, which could be raised and lowered by pulleys. The birds were driven into them. European fowlers also made elaborate liming stations, known in France as the *pipée,* where trees were trimmed to make perches, on which limed twigs were attached and the fowler concealed himself in a hut of brushwood. The name of such stations derived from the call used to draw birds into the catching area (Macpherson 1897).

Figure 9.2. *An Italian breccianella, or permanent bird trapping site. From MacPherson (1897).*

It is impossible to accurately assess the numbers of birds taken annually in these operations. MacPherson (1897) showed that there were marked annual variations in the numbers of many species caught, which largely reflected similar variations in the scale of their movements. Nor was every site equally successful. Attempts to calculate totals from the number of sites and samples of birds taken are always going to be very loose estimates at best, but some figures are given under the species headings below.

The operation of fixed sites was just one of the methods used for trapping birds. An unknown number of peasant bird-catchers in Europe, who lacked specific rights, set their nets and traps in areas of free land, where they would not be turned away (MacPherson 1897). A few examples of the ways they operated are given here; others will be found under the species headings below.

Italian peasants caught Goldfinches *Carduelis carduelis* by setting seeding heads of thistles, teasels or hemp in the catching area of their clap-nets, an arrangement similar to the 'throstle gardens' made in Heligoland (Gatke 1895). In Malta, Common Nightingales *Luscinia megarhynchos* (which were taken for food rather than as cagebirds), warblers and other small birds were often taken by being driven into nets spread over a low-growing carob tree (Wright 1864). Everywhere in southern Europe a favourite ploy in summer, which was theoretically prohibited, was to set nets or lime sticks by water and catch birds as they came to drink. Nearly all the methods described by MacPherson (1897) for catching small birds operated around the same theme – using food and decoys or call-birds to attract birds to the fowler, and a wide variety of artificial calls was also made to draw birds in.

Species taken

The passerines most widely sought for food were larks, especially the Skylark, thrushes, particularly Fieldfares and Song Thrushes *Turdus philomelos*, Northern Wheatears *Oenanthe oenanthe*, and some of the seed-eaters, particularly Ortolan Buntings *Emberiza hortulana* and Corn Buntings. These are dealt with in more detail under the species headings below.

Many other species were also taken. For example, Bolam (1912) remarked that he had known as many as 20 Snow Buntings *Plectrophenax nivalis* picked up dead at 'a single discharge of no.5 shot and have proved afterwards their excellent qualities upon the table'; he considered them as not inferior to Ortolans. Harvie-Brown & Buckley (1895) noted that Snow Buntings occurred in 'astonishing numbers on the braes of Glenlivet' in Moray, and were shot in large numbers there for eating. They quoted J. G. Phillips, who said that 'a long line of oats is laid down, called a *ghosk*, the great flights alight and feed, and the man, watching his opportunity, fires'. He recalled 118 being shot at one discharge, and remarked that the flesh was delicious. Snow Buntings were also killed for food in Caithness, but shooting them was not otherwise common in Scotland (Harvie-Brown & Buckley 1887). Nevertheless, Muirhead (1889) noted that winter flocks in Berwickshire were

> very erratic in their movements … according to the weather. They feed of various grass seeds, which they find on the moors and leas, and likewise on the waste corn in the stubble fields; but when severe weather sets in they are sometimes driven to visit stackyards.

In such sites they were vulnerable to people hunting seed-eaters, as described above.

In Europe Pine Grosbeaks *Pinicola enucleator* and Waxwings were also trapped in large numbers during their periodic irruptions, and females of many songbirds trapped for the cagebird trade were also sold for food, at 3d to 4d per dozen in the 18th century.

Shrikes were not only trapped for use as lures in trapping falcons; they were also taken as pets and used for hawking small birds in France, China and Japan. They were highly regarded as table birds in Italy, and Macpherson (1897) quoted some figures for the numbers received by one Florentine poulterer from Fano in 1886. These amounted to 378 shrikes on August 25th, 810 on September 1st, 400 shrikes and Calandra Larks on 3rd, 300 shrikes and 700 Calandra Larks *Melanocorypha calandra* over the 4th and 5th, and 176 shrikes on the 15th, after which trapping ceased. The species involved were Lesser Grey Shrike *Lanius minor,* Woodchat Shrike *Lanius senator,* and Red-backed Shrike *Lanius collurio.*

Shrikes are still caught for food round the Mediterranean today. Woodchat Shrikes are hunted in summer and on migration on a sufficient scale in Italy, southern and eastern Spain and North Africa to probably have an impact on the population (Tucker *et al.* 1994). Tucker *et al.* also noted that shooting on autumn migration in Turkey and parts of the Middle East was a conservation threat to all shrike species. Goodman & Meininger (1989) show a photograph of a bunch of live passerines offered for sale in Rosetta market, many of which are clearly shrikes, and Norbert & Worfolk (1997) noted that large numbers of Red-backed Shrikes were still taken on passage in the Greek islands. They also recorded that the birds were taken in Turkey for use as lures to trap the Sparrowhawks *Accipiter nisus* used for Quail *Coturnix coturnix* hawking.

Starlings *Sturnus vulgaris* were widely taken in Europe, although Ray (1678) said of them that 'Stares [starlings] are not eaten in England by reason of the bitterness of their flesh; the Italians and other Outlandish people are not so squeamish'. Young Starlings were relished on Heligoland, and Gatke (1895) recorded that large numbers were shot there in early autumn, but the old birds, which passed through the island later, were ignored as dry and tough. Villagers in Scandinavia, in Italy and probably elsewhere in Europe placed artificial clay nest-sites, similar to the sparrow bottles used in England (see p. 156), under the eaves of their houses to encourage Starlings to nest and provide a harvest of young for the table (MacPherson 1897). Large numbers of Starlings were also taken in Germany and the Netherlands in nets set in reed-bed roosts; MacPherson recorded that some fowlers placed baskets baited with cherries in such roosts, and caught considerable numbers in that way.

A fruit diet was held to enhance the palatability of many passerines. One widely sought all around the Mediterranean was what was known as the *beccafigo,* the fig eater. This species was identified with the Blackcap *Sylvia atricapilla* by Ray (1678) and the Garden Warbler *Sylvia borin* by Wright (1864) and Lilford (1895) but the name was perhaps applied to other species of the genus *Sylvia,* or other small passerines feeding on figs and grapes in the autumn. Large numbers were taken in August and September, Wright, for example, noting that in Malta as many as 100 dozen Garden Warblers were sometimes brought into the markets at one time.

Macpherson (1897) also described the use of 'spider-nets', woven from fine silk and often dyed green, which, by his description, were little different in design and use to modern mist-nets. The early- to mid-19th century Essex naturalist Dr Allan Maclean was adept at using spider-nets, and in a letter to T. C. Heysham said 'I have taken all the warblers except the Wood Wren [Wood Warbler *Phylloscopus sibilatrix*], Grasshopper Warbler [*Locustella naevia*] and Dartford Warbler [*Sylvia undata*]; also Kingfishers [*Alcedo atthis*], Rock Larks [perhaps Rock Pipit *Anthus petrosus*?] and an endless number of other birds, and I have not the least doubt but I could take any birds except those which are constantly at the tops of high trees'. His nets for warblers were made from the finest knitting-silk and would hold birds up to the size of a Blackbird *Turdus merula* (Christy 1890).

Passerines are still hunted on a considerable scale around the Mediterranean today, although McCulloch *et al.* (1992) recorded a declining trend in hunting pressure with many species since 1980. However, rather sparse data suggested that substantial numbers were still taken in the eastern Mediterranean, particularly in Cyprus, where Magnin (1987) claimed that two million birds were taken for food annually in mist-nets and by liming, and Malta, where the estimate was around three million birds.

Skylarks and other larks

Three species of lark have been regularly exploited as food in Europe; the Skylark, the Woodlark *Lullula arborea* and, in southern Europe, the Calandra Lark *Melanocorypha calandra*. No doubt other species were taken (which have generally been recorded simply as 'larks'). By the 19th century the Woodlark was taken mainly as a cagebird in Britain, but Edward Topsell, writing in the 17th century, regarded it as superior on the table to Skylark (Harrison & Hoeniger 1972). Both the other species were valued as cagebirds but particularly as food, and Skylarks were taken for this purpose in prodigious numbers.

Although they do not appear particularly frequently in the archaeological record in Britain for the Roman or mediaeval periods (see Tables 1.1 and 1.2, p. 18/19), Skylarks were the most frequent passerine recorded in the mediaeval and early modern household accounts used in compiling Table 1.3 (p. 20) and the market price-lists given by Jones (1965) in Table 1.4 (p. 21). They were always regarded as something of a delicacy, and were eaten roasted or made into pies, puddings, pâté or stews; in the latter they might be flavoured with ginger and cinnamon (Muirhead 1889). Woolgar (1995) suggested that the pattern of accounting for larks in some mediaeval household accounts indicated that they were sometimes held in mews (see p. 25) until required, a point also made by Stone (2006). Skylarks were still sold in the London markets up to 1941, when the price was listed as four shillings (around £8 today) per dozen (Gladstone 1943).

In the mediaeval and early modern periods, the prices of Skylarks were controlled in many markets and gentry households; Table 9.1 summarises the prices recorded in various sources in pennies (d) per dozen. A 50% increase in prices is evident in the lists given for London for the 16th century by Jones (1965). Jones also showed that prices tended to be higher in the winter and early spring than in the autumn.

There was a substantial domestic and international trade in Skylarks. Mayhew (1861–62) recorded that an average of 313,000 larks passed through the game markets of Leadenhall and Newgate in London each year; for 1854, Yarrell & Newton (1874) put the figure at 400,000, with up to 30,000 being consigned at one time. Newton was told by one large firm of poulterers that the bulk of those sent to London were 'worthless' and sold at an average price of only 15d per dozen, which still valued the trade at £2,000 per year (some £200,000 today).

Year	Place	Price per dozen	Reference
1274	London	1d	Jones 1965
1393	York	1d	Nelson 1907
14th century	London	Average 3d	Jones 1965
1512	Northumberland household accounts	2d	Nelson 1907
1560	Hull	4d	Nelson 1907
16th century	London	Average 5³/₄d	Jones 1965
1612–1634	Naworth accounts	Average 2¹/₂d	MacPherson 1892
17th century	London	10d	Jones 1965
1807	London	48d	Gladstone 1943
1922	London	36d	Gladstone 1943
1941	London	48d	Gladstone 1943

Table 9.1. *Prices of Skylarks in old pence per dozen, from the late 13th century to the 20th.*

Mayhew (1861–62) noted that larks taken for food and sent to London in his day mainly came from Cambridgeshire, with a few from Bedfordshire, but in the 18th century the main English source was the area around Dunstable, where bird-catchers were able to operate in areas of open access on Dunstable Downs, and from whence Pennant (1776) recorded that 4,000 dozen larks per year were supplied alive to the London markets, and many were served as lark pie in Dunstable's inns (Trodd & Kramer 1991). Trodd & Kramer noted that the birds were caught from Michaelmas to February at night, by labourers using drag- or trammel-nets. On a good night, two men could take up to 15 dozen larks. Clap-nets were used there by day, but the numbers taken were far smaller. Yarrell & Newton (1874) noted that this trade had ceased in the 19th century, perhaps as a result of enclosure and increasingly strict game protection.

In the later 19th century, Brighton, where large numbers of larks were netted on the surrounding downs, was probably the main centre of the trade in Britain outside London. Walpole-Bond (1938) noted a substantial trade between Brighton and Dieppe in the 19th century and a note in the *Zoologist* for 1892 (p. 409), under the heading 'Exportation of Larks and Thrushes' recorded that

> *Quite a trade in Larks and Thrushes is carried on between Brighton and Paris, throughout the whole of the winter season, by wholesale netting on the Brighton Downs. From a dozen to twenty hampers full of these birds (averaging about 14 lbs [6.3kg] per hamper) are sent off daily to Paris alone, to say nothing of those which are retained for home consumption.*

Gray (1871) quoted official returns from Dieppe that showed that in the winter of 1867 to 1868, 1,255,500 Skylarks, valued at 56,497 francs, were imported into the town; 983,700 were consumed locally and 271,800 sent to other parts of France. Donald (2004) noted that in the 1830s, nearly a million Skylarks were sold annually in the Halles market in Paris.

However, Germany was the country in which most Skylarks were taken. Yarrell & Newton (1874) quoted tax returns from Leipzig, which indicated that an average of 5,184,000 were received annually there; 404,300 were supplied to the city in one month in 1720, and 500,000 were supplied in October 1824. The birds were mainly caught in open parts of Saxony, and even more were sent to Berlin, Hamburg and other cities and towns. This traffic in Germany had largely ceased in the late 19th century (Yarrell & Newton 1874).

Despite the scale of these depredations, the 19th and early 20th century avifaunas for Britain, at least, provide little evidence that large-scale trapping was then affecting native breeding populations. Bird-catchers, particularly in areas such as the Sussex Downs, were mainly exploiting the massive autumn and winter movements of Skylarks and other birds out of eastern, central and northern Europe, which were a prominent feature of the early literature of these species in Britain. Kearton (1902) recorded that during one such movement in January 1897, the Brighton bird-catchers took 1,000 dozen larks in one morning before midday.

By the late 19th century, the county avifaunas show that Skylark netting in England and Wales was largely confined to a shrinking area of extensive stretches of land with open access, such as the Sussex downland or extensive dune systems in northwest England. But Skylarks mainly wintered in stubble fields (Shrubb 2003), and Knox (1849) noted that the main way of taking them in his day was to drag these stubbles at night with a drag- or trammel-net, a method also used to take partridges. The same method was used in Bedfordshire in the 18th century (Trodd & Kramer 1991). It is a reasonable assumption, therefore, that netting Skylarks was increasingly restricted from the mid-19th century by the scale of enclosure and the rise of close preservation of game. The preservation of large stocks of partridges and the nocturnal netting of Skylarks were incompatible, and the latter was hardly likely to be tolerated by landowners or their gamekeepers, who sowed their fields with cut thorns to entangle the nets. Booth (1881–87) designed and had manufactured a steel thistle with sharpened points for this purpose, which could be driven into the ground and was guaranteed to rip up nets – and, indeed, poachers' legs, if they walked into it.

There is little evidence that larks were taken on any scale in Scotland, although they figure in the household accounts of James V in the early 16th century (Bourne 2006). Harvie-Brown & Buckley (1892) noted that in some districts close to Glasgow large numbers were trapped and sent to Glasgow market; the same authors in 1895 recorded that Skylarks were declining in parts of Moray 'because of partridge netting at night'. Otherwise, the Scottish avifaunas for the late 19th and early 20th centuries do not record this practice. Nor do Ussher & Warren (1900) mention it occurring in Ireland.

Fowling methods

The scale of effort devoted to catching Skylarks seems of particular interest. Besides the use of drag- or trammel-nets and clap-nets, the main ways of taking them were with long flight-nets in Germany (known as the *lerchenstreichen*), snares in Lancashire, and the low-bell (see below). They were also taken by falconers and, in the later 19th century, shot over the 'lark-glass', also known as a daring-glass or a twirler. Essentially this was a length of wood of trapezoid section, inset with pieces of mirror and mounted on a spindle which could be spun by a long cord; early designs in England were painted red. The Lancashire bird-catchers also attached a piece of red cloth to it (Mitchell 1885). When spun, the flashing of the mirrors attracted larks to flutter round it.

The forms and methods of use of trammel-nets and clap-nets were summarised on page 35, and their use in the pursuit of larks showed no unusual features except that a lark-glass was often set in the middle of the clap-nets to draw the birds to within catching range. In the 17th century, Aubrey (1969) described this procedure:

> *We have great plenty of larkes, and very good ones, especially in Colerne [in Wiltshire] fields and those parts adjoyning to Coteswold. They take them by alluring them with a dareing-glasse, which is whirled about in a sun-shining day, and the larkes are pleased at it, and strike at it, as at a sheepe's eye, and at that time the net is drawn over them.*

Mitchell (1885) quoted William Blundell of Crosby, writing in 1620, who said that 'when the sun doth not shine, a fox tail pulled up within the compass of your net will make the larks strike at it as if it were a weasel'. Mitchell noted that 8–10 dozen larks per day could be taken with one clap-net.

The lark-glass was used in the 19th century to attract larks to the gun, and Knox (1849) left a detailed description of what went on with it in the Brighton area (see Donald 2004). But it seems not to have been practiced in many other areas, and Knox's description suggests why; it was unlikely to be tolerated by landowners or game-preservers on their land. It was, however, widely practiced in France, and Donald illustrated it well. Interestingly, Lilford (1895) noted that his resident Skylarks ignored the lark-glass completely and only migrants were attracted.

The low-bell and the lerchenstreichen

Low-belling also involved the use of a trammel-net, combined with a light and a bell, often a cow bell. According to MacPherson (1897) the word 'low' derived in this instance from an old dialect word *lowe*, and refers to the light not the bell. The right to pursue this sport was valuable, and, unlike the right to take gamebirds, was attached to the ownership of the land, so it could be traded between persons. Various descriptions of low-belling can be found in the early literature; Ray (1678) noted it as chiefly of use in champain country, *i.e.* open field districts, from the end of October until the end of March, and described the method as follows:

> *About eight of the clock at night, the air being mild, take your Low-bell, of such size as a man may well carry in one hand, having a deep, hollow, and sad sound; and with it a net of small mesh at least 20 yards deep, and so broad as to cover five or six ordinary lands [a land was presumably the ridge between two furrows in the ridge-and-furrow pattern of the open fields; according to Rackham (1986) this averaged about 11 yards (10m) across], or more as you have company to carry it: and go into a stubble field (a wheat stubble is the best). He that carries the bell must go foremost, and toll it as he goeth along as solemnly as may be, letting it but now and then knock on both sides. Then shall follow the net borne up at each corner, and on each side. Another must carry a pan of live coals, but not blazing. At these, having pitched your nest where you think any game is, you must light bundles of Hay, Straw or Stubble, or else Links or Torches, and with noises and poles beat up all the Birds under the Net, that they may rise and entangle themselves in it, and you may take them at pleasure. Which done extinguish your Lights, and proceeding to another place, do as before.*

The practice disappeared from the record in England in the 18th century, it having been prohibited by the Game Act of 1671 (Munsche 1981), largely because partridges were also taken by it. But it persisted in southern Europe. In Italy the method involved a long-handled hand-net known as the *lanciatoia*, rather than the trammel-net (Figure 9.3). It was also illegal there, a prohibition rarely regarded. Chapman & Buck (1893) also recorded its use in Spain, saying

> *They employ the cencerro, or cattle-bell and dark lantern. As most cattle carry the cencerro around their necks, the sound of the bells at close quarters by night causes no alarm to the birds. The bird-catcher, with his bright candle gleaming before its reflector and the cattle-bell jingling at his wrist, prowls nightly over the stubbles and wastes in search of roosting birds. Any number of bewildered victims can thus be gathered, for larks and such-like birds fall into a helpless state of panic when once focussed in the bright rays of the lantern.*

Calandra Larks were often taken with this method in both Italy and Spain.

The use of the lerchenstreichen was confined to parts of Germany, and required a substantial investment of capital and commitment of manpower. It consisted of a series of long nets, around 300m long and two metres high, across an open area of stubble. Up to a dozen such lines, set four to six metres apart, constituted one wall of netting; more than one wall might be set at a time. The season for operating this system ran from mid-September to

Figure 9.3. *Catching larks with a light and the hand-net known as a lanciatoia. From Ray (1678).*

the end of October, and 20–30 men were needed to drive the larks into the nets. This was done after sunset so the birds would not see the nets, and the drivers dragged the stubble with a line. Often the bag was up to 30 dozen, but at well-favoured sites up to 1,500 birds could be taken in an evening (MacPherson 1897).

Several other methods were used to take larks in a more local way. Mitchell (1885) recorded that the main means of catching them in Lancashire in his day was with lines of snares known as pantles (see p. 127) and thousands were caught in the sand dunes and on the mosses with these devices, and sent to market. MacPherson (1897) recorded that larks were sometimes taken in northern England and in Germany by stretching a long line across a field from which dangled a series of limed threads, and driving the birds gently up to it.

Falconry

Falcons were also used to take Skylarks. The Le Strange household accounts recorded 14 larks killed with a Hobby *Falco subbuteo* in June 1533, and 12 more a fortnight later (Gurney 1921). These entries should not perhaps be interpreted literally. Salvin & Brodrick (1855) noted that Hobbies were easily trained to 'wait on', and that the presence of a Hobby overhead caused larks to freeze; fowlers took advantage of this to catch them with a light trammel-net while a Hobby 'waited on' overhead. This was known as 'daring'; it was practiced in the summer and early autumn, when the birds caught must have been local residents. Merlins *Falco columbarius* were more usually flown at larks, but this sport had far more to do with the spectacle of the flight than securing larks for the table. Anyone who has watched Merlins hunting larks in the wild will readily appreciate this.

Larks remain legitimate quarry in Mediterranean Europe and, although the hunting of passerines has generally declined since 1980, that of Skylarks (and Song Thrushes) has increased, perhaps to compensate (McCulloch *et al.* 1992). While there are apparently no indications that continued trapping is affecting breeding populations of Skylarks, Yeatman (1976) recorded that hunting Calandra Larks in France was contributing to that species' decline there.

The Wheatear

As with the Skylark, the Northern Wheatear was long regarded as a delicacy in England. It did not, however, appear in those early household accounts or market lists I have examined, perhaps because it was subsumed into the category of 'small birds'. From the 17th century, at least, the trapping and eating of Wheatears was regularly discussed and it was one of the species that the London poulterer Adam Shewring was fattening for his customers in the late 17th century (Thirsk 2007).

A Sussex trade

Trapping Wheatears was apparently always a valuable secondary trade of the shepherds on the Sussex Downs; there are few records of it being practiced anywhere else in Britain. A record in a late 17th-century manuscript edited by Brentnall (1947) suggested that it may then have been practiced in Wiltshire, while Pulteney, writing in 1813, recorded it on the Isle of Portland in Dorset (in Mansell-Pleydell 1888), but Mansell-Pleydell did not record it as continuing later in the 19th century. Although Kelsall & Munn (1905) infer that Wheatears were taken in the New Forest without saying how, none of the other 19th-century county avifaunas (except those for Sussex) say anything about it. Similarly Ray (1678), Pennant (1776), White (1789), Montagu (1833) and Yarrell (1845) all record Wheatear-trapping only as a speciality of the East Sussex downland.

The method used by the Sussex shepherds to trap Wheatears was unchanged from the 17th to the late 19th century, and was described by Hudson (1900). He wrote

In July the shepherds made their 'coops', as the traps were called – a T-shaped trench about fourteen inches long, over which the two long narrow sods cut neatly out of the turf were adjusted, grass downwards. A small opening was left at the end for ingress, and there was room in the passage for the birds to pass through toward the chinks of light coming from the two ends of the cross passage. At the inner end of the passage a horsehair springe was set, by which the bird was caught by the neck as it passed in, but the noose did not as a rule strangle the bird.

Hudson noted that, on some of the high Downs between Rottingdean and Beachy Head, the shepherds made so many coops, set close together, that the downs in some places looked as if they had been ploughed. When the season finished the sods were replaced, roots down. According to Yarrell (1845) it was not unusual for a shepherd and his lad to look after as many as 700 of these traps, and Markwick (1795) noted that as many as 84 dozen Wheatears could be taken by one shepherd in a day, although the usual number was three or four dozen.

The season for taking them was late July to late September and the birds taken were mainly autumn passage migrants, recorded moving through the area in successive waves. It is quite evident from the reports in the literature that this autumn movement was on a far greater scale than anything we see today. Yarrell (1845) noted that it was customary to dress them by dozens at the inns of the numerous watering places along the Sussex coast, and Pennant (1776) recorded that about 1,840 dozen were taken annually around Eastbourne alone, with the trade then practiced along the length of the higher Downs from Beachy Head to Brighton. It may have been carried on further west in the 17th century, as Ray (1678) described it as from the whole length of the Sussex Downs. Traditionally the birds were known as 'Ortolans' in Sussex (Hudson 1900), which may have caused some confusion in earlier records.

The shepherds earned a good deal of money from this trade. In the 18th century, Wheatears were worth 6d per dozen to them, and Pulteney (in Mansell-Pleydell 1888) recorded that one trader in Weymouth paid £30 to a trapper from Portland in 1794 for Wheatears at a shilling per dozen, and was offered 50 dozen more than could be disposed of. The price rose to 1/6d per dozen in the early 19th century and 3/6d (about £17.50 today) at its end. Shepherds could each earn £12–£14 (£1,080–£1,260 today) per season, a sum equivalent to nearly half their annual wage (Shrubb 2003), and sometimes took as much as £30–£50 (£4,500 today) (Hudson 1900).

Hudson (1900) noted that the terms of employment of the shepherds changed in the second half of the 19th century. Before then the shepherd was paid partly in kind – so many lambs at lambing time and these, when grown, could be run with the flock he managed to a limit of 20–25 head, giving him a stake in the good management of the flock. However, this was changed to straight money wages in the late 19th century, and employers then stopped the shepherds from trapping Wheatears, on the grounds that it distracted them from their proper work, which had not been a problem before. But Booth (1881–87) gave another reason, noting that the area of sheep pasture was declining in favour of arable land, and flockmasters perceived the encroachment of large numbers of 'coops' to be against their interests. He recorded that 1878 was the last season in which poulterers received a supply from the shepherds, and what reached the market thereafter were taken by bird-catchers in clap-nets.

So far as I am able to tell, the method used by the Sussex shepherds to trap Wheatears was unique to the Sussex Downs. In Italy MacPherson (1897) noted that Wheatears were mainly taken in cage-traps or by lime sticks set around a decoy Little Owl *Athene noctua*. In Germany lime sticks were also used, set in rows across a field with the birds simply driven gently into them, a method that seems curiously inefficient. In Heligoland they were taken in large numbers in clap-nets set along a cliff edge, and served as delicacies to visitors who came for the sea-bathing (Gatke 1895). Considerable numbers could be caught, Gatke recording up to 10 score per day in a large movement.

Thrushes

After the larks, thrushes were probably the group of passerines most commonly trapped for food in the past. Over much of Europe the main season for taking them was the period of the autumn migration. Thrush-catching was well-known in both classical Greece and Italy, and MacPherson (1897) observed that the caterers who contracted for the kitchens of the rich in the Roman Empire made an extensive business of fattening thrushes in thrush houses or mews, each of which could fatten up to 5,000 thrushes in a season. It was a business that paid better than managing a good farm. The birds were confined in a dark room and encouraged to gorge themselves until their bodies became loaded with fat. Pennant (1776) noted that the birds were fed on food such as mixed fig and bread crumbs; fig-fed thrushes were regarded as a great delicacy. In the 17th century at least one London poulterer was fattening Blackbirds, Fieldfares and other thrushes for his customers (Thirsk 2007), who commented that, since his methods were known to earlier writers, some domestic households were most probably doing the same.

Thrushes, perhaps particularly Blackbirds, appear regularly in mediaeval and early modern household accounts and market lists and are probably under-recorded, some being hidden in the categories 'great' or 'little' birds that also appear with some frequency. Walpole-Bond (1938) noted a considerable trade in thrushes as well as Skylarks between Brighton and Dieppe in the late 19th century (see p. 162), caught in clap-nets by the Brighton bird-catchers. Thrushes were still being sold in English markets, mainly as shot birds, in the early 20th century, when Patterson (1905) drew attention to the quantities to be found on game dealers' stalls in Yarmouth. Hawker's diary for February 2nd 1831 records a massive influx of Fieldfares in a snowstorm,

> *so tame that you might have kept firing from morning till night, though I found it impossible to get more than five at a shot. After killing as many as I wanted, without even moving from the hedge I took shelter under, I weathered the storm … It was quite laughable, when the storm ceased this afternoon, to see and hear the levy en masse of tag-rag popgunners blazing away at the Fieldfares.*

Simon (1952) gave a 19th-century recipe for Blackbird pie, and Thirsk (2007) noted that the breasts of thrushes and Blackbirds were still relished as food in north Lincolnshire in 1915, baked in pies with hard-boiled eggs and bacon.

Snaring

The following notes on thrush-snaring and netting in Europe are drawn largely from MacPherson (1897). Comparatively little appears to be recorded on the subject in Britain before the 19th century, although Willughby and Ray devoted a whole section of the *Ornithology* to the general trapping of passerines (Ray 1678). In Europe the principal ways of taking thrushes were by snares and nets, although some were taken with bird lime, particularly in southern France, and in Germany Fieldfares were apparently killed in large numbers with slings. Whilst this seems slightly improbable, what was meant may have been the crossbow known as a *ballista* (see p. 34).

Thrushes were snared throughout Europe during the autumn migration. They were trapped in particularly large numbers in vineyards, with snares baited with grapes (Brusewitz 1969). Immense numbers were also snared annually in olive groves in Sardinia between the end of October and February. Large numbers of Blackbirds were taken in Corsica until recently, for Ehrlich *et al.* (1994) reported that an annual production of five tonnes of Blackbird pâté had only just ceased there. Where vineyards and olives were not available, snares were attached to suitable trees and baited with wild berries (Figure 9.4).

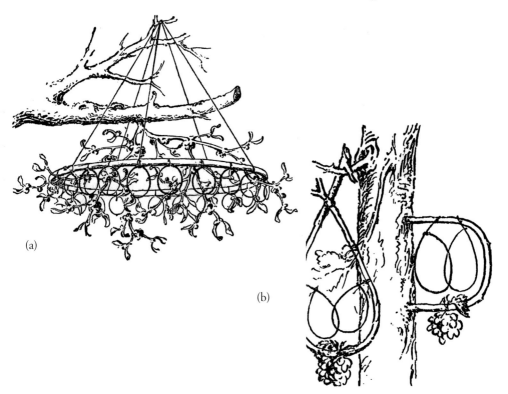

(a)

(b)

Figure 9.4. *A French Mistle Thrush-snare (a), and a Norwegian thrush-snare (b). From MacPherson (1897).*

Particular baits were used for particular species, mistletoe berries for Mistle Thrushes *Turdus viscivorus*, service berries for Waxwings (regarded as a thrush in the past), and rowan, elder, bird cherry or juniper for other thrushes. The Germans had a particular liking for thrushes that had been feeding on juniper berries, and MacPherson quoted William Turner's *A New Herbal*, published in 1551, that juniper

> *groweth in Germany in many places in greate plentye, but in no place in greater than a little from Bon [Bonn], where as at the time of year the feldefares fede only of Junipers berries, the people eat the feldefares undrawen with guttes and all because they are full of the berries of Juniper.*

The demand for rowan berries as bait for snares led to a brisk trade in harvesting the wild fruit and selling it, for five francs a bushel; in poor berry-years the price could rise to two pounds. The existence of such a trade is a clear indication of the scale on which such trapping was carried out.

Fowlers in Belgium and Luxembourg hired what were known as a *tenderie* for the specific purpose of snaring thrushes. This consisted of 4–5 acres (1.6–2 ha) of underwood from 3–5 years old; the rent amounted to around 30 shillings per year. The same spot was rented by one man in successive years. Up to 150 birds could be taken in a day, which sold for between 5d and 11d per dozen.

Methods of snaring varied little across Europe, although the Poles used a type of springe (see p. 130) rather than a hanging snare. Much of the catch was sold to dealers, who sent it on to major urban markets. By the late 19th century thrush-snaring had been made illegal in Denmark, and attempts were made to restrict it in Germany by protecting species that bred, though these met with little success.

There are few figures available with which to measure the scale of this activity but figures given by MacPherson (1897) for one district in the Rhine provinces of Germany for the early 17th century showed that between 1611 and 1632, 14,874 thrushes were taken in snares, of which 35% were Song Thrushes and 50% Redwings *Turdus iliacus*. Netting took similar numbers. One good netting station in Germany is recorded as having taken a total of 7,222 thrushes between 1819 and 1832, of which 72% were Redwings, 14% Fieldfares and 14% Song Thrushes. These figures from Germany make an interesting contrast with bags from Italy, where by far the greatest number of thrushes taken were Song Thrushes and Redwings; Fieldfares often hardly figured.

Netting

Thrushes were mainly netted in large clap-nets in Germany, often set up in permanent stations in woodland; areas where there were extensive areas of juniper were particularly favoured. Large numbers of Song Thrushes were taken on Heligoland in artificial *troossel-goard* or thrush gardens. These were made especially to trap thrushes and were described by Gatke (1895) as a space about 20 x 8 feet (8 x 2.5m), surrounded by bushes 10 feet (3m) high, simply stuck in the ground fairly closely together and surrounded by netting, with one side left open. They were sited where they were obvious to migrating thrushes from a distance, and proved an irresistible attraction to the birds in a largely bare autumn landscape. Once in the bushes the birds were driven into the nets. There were about 20 of these devices on the island and it was by no means rare to catch several hundred birds in a single morning at one site; a thousand had been known. Gatke noted that the numbers taken declined considerably towards the end of the 19th century, with a decline in the scale of movements arising from a change in the prevailing wind patterns. A hundred in a day was then considered an exceptionally good catch.

Fixed sites such as the roccolo (see p. 172) were important for netting thrushes in Italy, but the Italians also used long nets, set around woods the thrushes used as roosts into which the birds were driven at dusk, and flight-nets. These were set across flight-lines, which the birds used regularly to descend from the hills to the plain. Ownership of such a site was valuable and was leased to the fowler. From at least the 16th century the Swiss also netted thrushes as they migrated through mountain passes. Their method was to cast stones and other missiles into the air with slings to induce the frightened birds to dive down into fixed nets standing about six feet (1.8m) high. Presumably this method was more efficient than it sounds.

MacPherson (1897) noted a 'vast diminution' in the numbers of thrushes being trapped in Italy in the late 19th century. He had no hesitation in ascribing this to the scale of snaring and netting that was pursued during the century. But the trapping and shooting of songbirds continues around the Mediterranean today and, like Skylarks, thrushes remain legitimate quarry in this region. Examination of long-term trends in hunting pressure there show that, contrary to the general declining trend in the hunting of passerines since 1980, the hunting of Song Thrushes has, like Skylarks, increased (McCulloch *et al.* 1992). The birds taken are mainly migrants from northern and eastern Europe, but these migratory populations have not shown signs of decline as a result (Tomialojc 1997).

The Ortolan Bunting

The Ortolan Bunting has been prized as a delicacy for the table in Europe since classical times, when it was fattened in special houses for the tables of Roman epicures. According to Edward Topsell, writing in the 17th century, the Romans knew them as *miliariae* (Harrison & Hoeniger 1972), a name which has, at various times, been incorporated into the scientific name for the Corn Bunting *Emberiza calandra* (e.g. *Emberiza milaria, Milaria calandra*), a bird which had an almost equal reputation for the table. In Topsell's day Ortolans were known in Italy as *cynchrams* or *cenchrams*. He wrote

> *The cenchrams take their name of millet seede whereupon they feede and men nourishe them in their fowleryes or birde-howses, which being fat they sell very deare. There was a lawe that none but senators and princes might eat thereof, for they were desired above pertriges, and better esteemed amonge nobles yielding a better rellishe, a sweeter and easier concoction and especially in the winter; for which cause they did sett them upp a fattinge, for they are a most delicate and wholesome foode.*

From at least the 16th century, and presumably earlier, Ortolans were taken in large numbers, usually in clap-nets, in northern Italy, southwest France, Switzerland and many parts of eastern Europe. But, except as rare (and probably overlooked) vagrants, they were unknown in Britain before 1837, when they were first imported into the London markets from Prussia (Yarrell & Newton 1882). Thereafter throughout the 19th century, considerable numbers were imported into London every spring, first from Prussia and latterly from the

Netherlands, although Yarrell & Newton suggested that the source of these birds remained Germany. They were presumably imported live and fattened in London; Yarrell & Newton discussed the possibility that vagrants reported in Britain around this time were escapes. Some no doubt were, as more than half the 19th- and early-20th century records were concentrated in London and neighbouring counties. Bucknill (1900) noted that 'in recent years' their place as table delicacies had been largely taken by Quail, which fits well with the marked increase in the import of Quail in the later 19th century (see p. 107).

Fattening Ortolans

The methods involved in the ancient practice of fattening Ortolan Buntings seem not to have varied for centuries. Thus, in Italy, they were kept in darkened rooms, with just enough light to enable the birds to eat and drink. Fresh water was constantly supplied, and the birds were fed liberally on millet seed. The room was furnished with bushes and perches where the birds could rest. Birds caught in April were ready for consumption in June. After killing and plucking, the bodies were packed in boxes filled with meal or bran, in which they would keep for up to three weeks, and dispatched to customers all over Italy (MacPherson 1897). Bewick (1826) recorded that Yellowhammers *Emberiza citrinella* were frequently fattened like Ortolans in Italy, and were considered very good eating.

Ortolans are still trapped in large numbers in southwest France, a circumstance probably contributing to their present decline in Europe. Claessens (1992) estimated that an average of 50,000 birds were taken per year in this region by some 1,000 bird catchers. This represented a figure two or three times greater than the French breeding population of the species, and many of these birds were migrants from further north and east. Fattening Ortolans still continues in this region, although banned elsewhere in France.

Figure 9.5. *The exterior of an Italian roccolo, another type of permanent bird-trapping site, a smaller version of the breccianella. The man on the ladder gives an idea of scale. From MacPherson (1897).*

CHAPTER 10
Cagebirds and collecting

The keeping of birds as pets is an ancient practice. The Romans kept large aviaries, and similar housing for birds, particularly for Goldfinches *Carduelis carduelis*, Linnets *Carduelis cannabina*, Greenfinches *Carduelis chloris*, Chaffinches *Fringilla coelebs* and Serins *Serinus serinus,* was also fashionable with the monarchs and nobility of 15th-century France (Gurney 1921, Yapp 1982). Yapp also discussed evidence that keeping mammals and birds as pets was common among the leisured classes in England in the later mediaeval period, citing regulations prohibiting members of religious orders from keeping them from the mid-13th century. Members of Magdalen College, Oxford were forbidden to keep thrushes and other singing birds in 1459, and other Oxford colleges extended this ban to include Song Thrushes *Turdus philomelos*, Nightingales *Luscinia megarhynchos*, Starlings *Sturnus vulgaris* and Blackbirds *Turdus merula* in the early 16th century. As Yapp pointed out, those in authority do not seek to forbid such activities if people don't practise them. Yapp could find no evidence of the existence of extensive aviaries in England at this period and, in Yapp (1981), drew attention to an interesting difference in the way birds were shown in mediaeval French and English manuscripts. They were often shown in flight in French manuscripts, as artists would have seen them flying in aviaries, but hardly ever in English ones, as artists would have been most likely to have seen them in individual cages, and sometimes painted them so.

Most of the species mentioned above were native to western Europe, but parrots were regularly imported into Europe from all parts of the world by the 15th century and the Lisle Letters include references to parrots and parakeets being sent as presents in the 1530s (Greenway 1967, Bourne 1999b). Yapp (1982) noted that Ringed-necked Parakeets *Psittacula krameri* are illustrated sufficiently frequently in mediaeval manuscripts to suggest that they were commonly imported from the 13th century, and they may well have been imported much earlier, for they appear in mosaics of the 6th century at Ravenna. In Britain, a green parrot is shown on a table in a portrait of the 10th Lord Cobham and his family, painted in 1567; his son also sits at the table, with a tame Goldfinch on his hand. The importation of exotic animals as gifts to rulers was also a well-established tradition, and included birds, such as pelicans and Crowned Cranes *Balearica pavonina* (see p. 67).

The birds most favoured as cagebirds were those of bright colour and/or pleasing song. In Britain Woodlarks *Lullula arborea*, Skylarks *Alauda arvensis*, Nightingales, Linnets, Goldfinches and Bullfinches *Pyrrhula pyrrhula* were particularly sought, although Swaine (1982) quoted one old bird-catcher in Gloucestershire, who said that 'Londoners would buy anything with a beak and a tail'. Yarrell & Newton (1882) noted that the Chaffinch was one of the most highly valued cagebirds on the continent. Ratcliffe (1997) drew attention to the popularity of Ravens *Corvus corax* as pets in Victorian times, which continued an old Roman tradition. They were favoured for their readiness to accept captivity, engaging behaviour and ability to mimic the human voice. There was a regular trade in young birds through Leadenhall market, at 10 to 15 shillings each (around £47.50 to £71.25 today).

London's bird-catchers

Daines Barrington, in Pennant (1776), gave an interesting account of the activities and methods of the East London bird-catchers in the 18th century, noting that, in the suburbs and particularly about Shoreditch, weavers and other tradesmen got their livelihoods in spring and autumn by catching birds. They operated during the migration periods in late September to early November and again in March, using clap-nets and call-birds and decoys in pursuit of pipits, Woodlarks, Linnets, Goldfinches, Chaffinches, Greenfinches and other birds; the Linnet was the species most often caught. Pipits and females of all species caught were killed immediately, and sold for food at 3d to 4d per dozen.

Because of the amount of gear they carried, the bird-catchers operated within a radius of three or four miles from home, an area that in the 18th century would have included a great deal of common and heathland. They regularly carried call-birds and decoys of several species in little cages to place around their nets; Barrington listed Linnets, Goldfinches, Greenfinches, a Woodlark, a Redpoll *Carduelis cabaret*, a Yellowhammer *Emberiza citrinella*, an Aberdevine (Siskin, *Carduelis spinus*) and perhaps a Bullfinch. They also carried what were known as flur-birds, birds attached to a moveable perch, which was raised and lowered into the catching area as possible targets approached. Considerable sums were sometimes given for a single song bird, and Barrington noted that 'the greatest sum we have heard of was five guineas for a Chaffinch that had a particular and uncommon note, under which it was intended to train others'. Bullfinches were eagerly sought. They sold well, being readily taught to whistle tunes. In Somerset, many young Bullfinches were taken from the nest and reared as songbirds, the same melody being continuously whistled to them in a darkened cage, when in a short time they learned to reproduce the air perfectly. Church bell tunes were always favourites (Lewis 1952).

In the 18th century Barrington recorded that this trade was unique to the area of London he indicated, there being no considerable sale for singing birds except in the metropolis. But this was certainly not true as the 19th century progressed. The 19th and early 20th avifaunas leave the strong impression that there was a marked increase in the demand for cagebirds from

Figure 10.1. *Clap-nets in use in Lancashire in the 19th century. From Mitchell 1885.*

the second quarter of the 19th century; Glegg (1935) dated it from *c.*1838 in Middlesex. It seems to have emerged with the growth of the great industrial conurbations of London, the Midlands and northwest England. It is tempting to suggest that the demand for cagebirds formed a link, even if tenuous, with peoples' rural past, which was certainly fresh in the memory of many. The Goldfinch was a particular favourite, of which Newton (1896) noted that 'its docility and ready attachment to its master or mistress make up for any defect in its vocal powers'. It was also relatively long-lived. Yarrell & Newton (1876–82) noted that this finch had been known to live for 10 years in captivity, and Mayhew (1861–62) suggested that they frequently lived for 15 or 16 years, and recorded one instance of 23 years.

London remained an important centre of the trade, and Mayhew (1861–62) recorded that about 200 bird-catchers were then working around London, operating wherever there were open fields, plains or commons. As in the 18th century, they used clap-nets to make their catch, and decoys to entice the birds within range (Figure 10.1). The species principally taken are listed in Table 10.1, together with the numbers caught annually, details of the prices they made and of mortality after catching. This was often high because of the overcrowded and dirty conditions in which the birds were usually kept in the bird shops. In addition to those listed, small numbers of Blackcaps *Sylvia atricapilla*, Jackdaws *Corvus monedula*, Magpies *Pica pica*, Starlings and Redpolls were also caught, and Harting (1866) recorded that Tree Sparrows *Passer montanus*, Brambling *Fringilla montifringilla* and Twite *Carduelis flavirostris* were sometimes taken at Kingsbury Reservoir. Mayhew (1861–62) noted that there was a considerable sale of House Sparrows *Passer domesticus* as pets to children. A more esoteric branch of the bird trade pursued by a few street-sellers was the gathering and sale of birds' nests. They sold for a few pence each; this can never have been a very profitable occupation.

Species	Average number caught annually	Mortality in early captivity	Prices for adult birds	Prices asked by street sellers
Larks[1]	*c.* 60,000	not recorded	2/6d	6d to 8d
Robin	3000 or fewer	*c.* 33%	up to £1	1/-
Nightingale	not recorded	>50%	10/- to £1	2/-
Blackbird	*c.* 28,000[2]	*c.* 25%	10/- to £2	2/6d to 3/-
Song Thrush	*c.* 35,000[2]	*c.* 25%	10/- to £2	2/6d to 3/-
Chaffinch	*c.* 15,000	*c.* 33%	2/6d to 3/-	1/-
Greenfinch	*c.* 7,500	not recorded	2/6d	2d
Goldfinch	*c.* 70,000	*c.* 30%	not recorded	6d to 1/-
Linnet	*c.* 70,000	*c.* 50%	1/- to 2/6d	3d to 4d
Bullfinch	*c.* 30,000[3]	*c.* 33%	2/6d to 3/-	1/-
Totals	*c.* 318,500	*c.* 35%		

Table 10.1. *Numbers of birds caught per year by the London bird-catchers, their mortality rates after catching, and prices asked. Data from Mayhew 1861–62;* [1] *'larks' includes Skylarks, Woodlarks, tit-larks (pipits?) and mud-larks (?). Birds taken for the table are excluded;* [2] *'Blackbird' and 'Song Thrush' each includes a significant number of birds taken from the nest and hand-reared;* [3] *'Bullfinch' includes some birds sent from Norfolk and Leicestershire.*

There was also a significant sale of Canaries *Serinus canaria*, totalling *c.*16,000–17,000 birds annually, about half of which were imported from Germany and the Netherlands, and the rest bred in England, particularly by weavers in Leicester and Norwich. In the late 17th century, the main source to England had been from Germany (Ray 1678). Bewick (1826) recorded that 'four Tyrolese usually bring over to England about 1,600 and though they carry them on their backs 1,000 miles and pay £20 duty for such a number, they are enabled to sell them at five shillings apiece', making the trade worth some £40,000 in today's money.

Nightingales were caught at night, usually at some distance from London, in areas such as Epping. Beyond remarking that they were usually caught by the catcher imitating the bird's note. Mayhew does not say how they were taken. But Harting (1866) told of a retired keeper he knew, a skilled bird-catcher, who used at one time to rent a cottage 'for which he paid £10 per year … if there was what he called 'a good Nightingale season' he made more than enough to pay his rent by the capture and sale of these birds. In one season alone he caught 15 dozen, receiving 18 shillings a dozen for them in London' (a total of around £1,350 today). In Sussex, Kearton (1909) found that migrant Nightingales were taken in gardens in baited spring traps.

The Brighton bird-catchers

Another area of major importance for the cagebird trade was centred on Brighton. Booth (1878), discussing bird-catching on the Brighton Downs, noted that, at the first hint of cold weather, vast numbers of Blackbirds, Fieldfares, Redwings *Turdus iliacus* and larks would appear, luring out hundreds of bird-catchers. There was much competition for the best pitches, and trappers left the town at midnight to book their spots, sleeping rough under hedges and banks. Booth saw up to 200 clap-nets at work on a favourable day, and noted that up to 50 dozen birds were commonly taken, and that catches not infrequently reached 80 dozen.

Kearton (1909) recorded seeing seven sets of clap-nets spread in such close proximity on a favoured piece of ground that he thought they must be interfering with one another. He noted that the bird-catchers usually worked in pairs, one man attending the nets and the other beating the adjoining ground to flush birds to him. Call-birds as well as decoys were used to draw the birds in, the decoys being much the same species as listed by Barrington in the 18th century (see p. 174). Call-birds were valuable, Kearton noting that Goldfinches often cost £1 (£91 today) each and Linnets 10/- to 15/- (£45.50 – £68.00 today). Hens were usually killed when caught, and sold for food.

In periods of heavy movement, flight-nets were used rather than conventional clap-nets. These were single nets measuring 25 yards long by seven feet deep (23m x 2m), operated in the same way as clap-nets but pulled over to meet the birds in flight and take any flying low enough to be caught. A skilled operator could take up to 80 dozen larks in a day during a large passage (Kearton 1909). Kearton noted that birds caught on the Sussex Downs mainly went to London – cagebirds to dealers in areas such as Seven Dials, and birds for the table to Leadenhall Market. But there was also a significant export trade in thrushes and larks, with the birds sent to France.

Population effects

These examples are just a few samples to illustrate the scale of bird-trapping practised in the later 19th century, particularly for the cagebird trade. Trapping was widespread, and there are few county avifaunas of the 19th and early 20th centuries that do not discuss the activities of bird-catchers under one species or another. Nevertheless, although it must be remembered that detailed censuses were not available, there is surprisingly little evidence that trapping for the cagebird trade had a serious impact on wild populations before the 19th century, and examination of the 19th and early 20th century county avifaunas, most of which were published after 1870, shows that populations of such popular cagebirds as Blackbirds, Chaffinches and Greenfinches were then increasing, and those of Skylarks and Song Thrushes were regarded as stable in Britain; all were considered common or abundant throughout. The Linnet was declining but Holloway (1996) still showed it as common over much of the country, and the Bullfinch, which declined quite markedly in the 19th century, was increasing again with protection in the early 20th (e.g. Ticehurst 1909). Both Woodlark and Nightingale were declining, but with Woodlark, at least, it is unclear how much the decline stemmed from trapping, which Harting (1866) observed had done much to eliminate the bird around London, and how much from significant changes in the extent of its nesting habitats (Shrubb 2003).

The Goldfinch was the species most affected. Not much information is available on its status in the 18th and early 19th centuries beyond general statements that is was 'common'. But this finch may have been extremely abundant. Cobbett (1957), an accurate observer, in Gloucestershire in 1826 saw a flock he estimated to be 10,000 strong, on a half-mile of roadside thistle banks, and he remarked on several other very large flocks in the same area. But virtually all the 19th century and early 20th century avifaunas noted a serious decline, ascribed to habitat change combined with excessive trapping. Large numbers were taken during the migrations on the Sussex Downs. Gray & Hussey (1860) recorded annual totals of 400–500 dozen from the Worthing area, and up to 1,050 dozen in some years.

As the British populations declined, the shortfall required to meet demand was made up by imports, first from Germany and then Russia (Macpherson 1897). Macpherson recorded that one supplier in Algeciras, Spain, sent 10,000 Goldfinches and Greenfinches to England in November 1894 and offered further consignments of 250 birds at one shilling each delivered. But the species that Macpherson noted as most important in leading to a trade in birds between England and Russia was the northern race of the Bullfinch, the nominate *pyrrhula*, much sought for its large size and rich colour. Improvements in the speed and capacity of transport (steamships and railways) no doubt contributed to the scale of such imports, and the distances over which they could be brought.

Other captive birds

Ornamental wildfowl have been kept since at least the mid-17th century, when Charles II formed a collection in St James's Park, which included pelicans (that are still kept there). Some wildfowl, at least, were imported (Kear 1990). Such ornamental wildfowl are kept widely today and have sometimes given rise to feral populations that are perhaps less welcome, such as Canada Geese *Branta canadensis*.

More esoteric uses for captive birds are also described in the literature. Lapwings *Vanellus vanellus*, often ones that had been wounded, were kept in gardens, where they controlled slugs and snails (Shrubb 2007). They often became very tame, and Montagu (1833) tells the delightful story of one kept by a clergyman in his garden, which spent the winter nights in the chimney corner in his kitchen with his cat and his dog, returning to roost in the garden in the spring. Other species were pinioned and kept for this purpose; Thirsk (2007) mentioned gulls, and suggested the practice may have been imported from China.

Collecting

The obsession with collecting stuffed birds and eggs, which peaked in the Victorian and Edwardian periods, was aptly described by Nicholson (1926) as a 'leprosy laid upon birds'. Such collecting had a long history. Robert Plot, reporting a Hawfinch record for Staffordshire in 1686, wrote that it was 'now in the possession of the virtuous Madam Offley, a lady that has excellent artifice in preserving birds' (McAldowie 1893). Naturalists in Britain and in continental Europe, particularly France, were building collections of stuffed birds and exchanging skins in the early years of the 18th century, following the example of Francis Willughby and Sir Thomas Browne in the previous century (Allen 1976). Professional bird-trappers were already being employed to procure specimens at this time. Few of these collections survived inadequate early methods of preservation. Nevertheless, Bewick, in preparing his *British Birds*, acknowledged access to 57 collections, all formed before 1830, for subjects for his woodcuts.

Collecting supported a significant level of activity at auction. The first sale at auction of mounted birds that Chalmers-Hunt (1976) recorded in his study of natural history auctions from 1700 was at Christie's on June 6th 1771; the first including an egg-collection

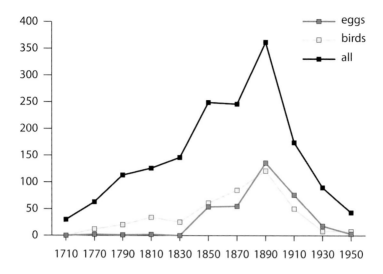

Figure 10.2. *The number of natural history auctions in the British Isles from 1710 to 1969 by twenty-year period, together with the number that included bird specimens and birds' eggs. Each twenty-year period is designated by the first year. The period from 1710 to 1769 is taken as one period because of the limited number of auctions. Data from Chalmers-Hunt 1976.*

he noted was at Gerard's between June 8th and 11th 1789. Altogether, Chalmers-Hunt traced a total of 17 such auctions up to the end of the 18th century, and a further 48 by 1830 (Figure 10.2). There is no doubt that collecting the skins and eggs of birds as a leisure pursuit increased markedly with increased prosperity in the second half of the 19th century, as Figure 10.2 shows, and continued at a significant level into the first half of the 20th century.

The scale of such collecting in the 19th and early 20th centuries is also illustrated by the number of collections that can be found listed in various sources, and the number of taxidermy businesses that operated in the same period. Table 10.2 lists the numbers of collections of stuffed birds I have found mentioned in the 19th and early 20th century county avifaunas for England and Wales by county, combined with the collections sold at auction for which Chalmers-Hunt (1976) was able to list localities, information that was by no means always available. Surprisingly few of the latter occur in the former source. A few additions were also found in Herriott (1968) and Frost (1987). In total they comprise 961–1,044 private collections and 118–120 institutional ones, in museums *etc*, some of which started as the private collections of benefactors. Some of these institutions appear to have been private enterprises but the assumption is that they were available for public view. The lists are incomplete; Chalmers-Hunt recorded the names of a further 126 owners of collections of birds and/or eggs without any indication of locality sold at auction in the period 1771–1939, and listed 268 anonymous consignments of birds or birds and eggs, and 231 anonymous consignments of eggs offered in the same period. The egg collections or consignments were all offered after 1850. How much duplication existed between these sales and between them and named offerings cannot now be ascertained.

These lists do not cover every county, nor were all these collections separately active at any time. There was a significant market between collectors, and many institutions received

County	Private	Institution	County	Private	Institution
Berks & Bucks	37	4	Norfolk	78	4–5
Breconshire	19	1	Northants	16–20	1
Cheshire	25–34	4	Northumbria	52–56	3
Cornwall	17–19	3	North Wales	37	1
Devon	59–63	5	Notts	18–19	2
Dorset	22–26	3	Oxfordshire	26	4
Essex	32–45	5	Pembrokeshire	12	1
Hampshire	28–36	9	Shropshire	24	3
Herefordshire	10–15	1	Staffordshire	14	2
Kent	47	9	Somerset	15	3
Lakeland	28	3	Suffolk	58–64	4
Lancashire	27–28	9–10	Surrey	35–39	2
Leicestershire	38–43	1	Sussex	30–35	7
Lincolnshire	5	?	Wiltshire	34–37	3
London	43–47	10	Yorkshire	78–76	11

Table 10.2. *Numbers of collections of stuffed specimens of birds recorded in England and Wales in the 19th and early 20th centuries (source: county avifaunas). Institutions are public museums, etc. Leicestershire includes Rutland, London includes Middlesex, 'Lakeland' comprises Cumberland and Westmorland, Northumbria comprises Northumberland and Durham (whose avifaunas overlap), North Wales comprises the counties of Anglesey, Caernarfon, Denbigh, Flint, Merioneth and Montgomery. The entry for Lincolnshire is incomplete, coming only from Cordeaux 1872.*

bequests of private collections. But these figures give some measure of the number of collectors active over the 19th and early 20th centuries. Many country houses, farmhouses and cottages also had cases of stuffed birds, which were not regarded as or, in the case of country houses, may not have been, formal collections; hence the range of figures given for some counties in the table.

Tubbs (1968) observed that, by the middle of the 19th century, ornithology and egg- and skin-collecting had become more or less synonymous, with the formation of collections to extend knowledge transmuting into collecting for collecting's sake. Collecting was a cult particularly of rarity and it was the cachet of 'British taken' backed by a published record which gave specimens particular value. And such specimens were valuable. The point made by Nicholson and Ferguson-Lees (1962) that prices of £5–£25 were quite commonly paid by wealthy collectors for rarities or unusual plumage variations is supported by the prices frequently quoted in the 19th-century county avifaunas. Pashley, the Cley taxidermist, obtained high prices; for example, £40 from E. Connop for Britain's first Pallas's Warbler *Phylloscopus proregulus*, shot in 1896 (Taylor *et al.* 1999). In an era when the value of money has been consistently debauched by governments, it is important to remember that £5 was regarded as a considerable sum of money in the 19th and early 20th centuries. It represented, for example, up to eight weeks' wages for an agricultural worker. Melling (2008), in his account of the Cheshire Kermadec Petrel in 1908, makes the point clearly. The £5 the Grosvenor Museum had to pay for the specimen was eventually raised by public subscription from 14 individuals and, on the basis of figures given by the Bank of England (see Appendix 5), the equivalent sum today would be £455 (all the prices quoted in this paragraph should be adjusted by the values shown in that Appendix; thus £40 in 1896 would be c. £3,640 today).

However, quoting prices is a fairly meaningless exercise in this context. The price a collector would be prepared to pay was always reflected in the rarity and condition of the specimen and how badly he wanted it; good provenance added to price, some collectors would always pay more than others, particularly to fill a gap in their collections, and some collectors were gullible. But there is no doubt that the increase in collecting activity indicated in Figure 10.1 was matched by significant increases in the prices obtained for good specimens at auction. This was well illustrated by Frost (1987) in his discussion of the prices obtained for three comparable Norfolk collections at auction in the 19th century, which showed that the average prices obtained for the rarer species, at least, increased eightfold between 1876 and the mid-1880s. Patterson (1905) noted that 'in recent years the value attached to rare and beautiful birds has been greatly enhanced by the eagerness displayed for collecting specimens, and the difficulty in procuring them in desirable conditions of plumage. Prices paid for local rarities, and birds in nuptial attire, have ruled exceedingly high, and as certain species become scarcer so will their money values correspondingly increase'. But it was probably more significant that a profitable and extensive trade existed to support collectors. This is evident from the number of taxidermists and dealers that were operating in the 19th and early 20th centuries. Sources such as the county avifaunas of the period, Herriott (1968), Frost (1987) and Morris (2010) list at least 501 dealers, taxidermists or bird-stuffers in England and Wales. Virtually every market or county town supported at least one such business. Nor are the lists in the county avifaunas complete. Harrop *et al.* (2012) quoted a list totalling 1,068 taxidermy businesses in England and Wales, which included a total of 297 for Yorkshire. The latter compares to a total of 32 listed by Nelson (1907) and Herriott (1968). Whether there were ever 297 taxidermists operating at one time in Yorkshire seems doubtful; many of these businesses were quite short-lived. It is also probable that authors of county avifaunas, such as Nelson, tended to work with a limited number of taxidermists operating in their areas, who they had found to be reliable sources of information. Frost (1987) noted that the London trade directories recorded something like 350 taxidermy businesses there, but no work of the vast majority has been recorded and I have only included here firms that were stated to have worked with birds; some specialised on mammals or fish, for example.

Many provincial businesses were small, and taxidermy was practiced alongside other trades. Hairdressing was quite common (see Plate 7) and tobacconists, furriers, cabinet-makers, antique-dealers, photographers, fishing-tackle makers, picture-framers and stationers are all recorded, whilst Samuel Clarke of Tuddenham in Suffolk was the village policeman, and Williams of Barrow was a blacksmith. Many of the more important taxidermists also amassed considerable collections of their own.

Provenance

Provenance – proper details of the source, place and circumstances of the record of the bird – was always important. One of the problems with what are known as the 'Hastings Rarities' (see Nicholson & Ferguson-Lees 1962) was that many of the records had no proper provenance, as the names of those procuring the birds were suppressed, on the grounds that there was no wish to get these men into trouble for breaking the law. It is not at all clear that they were doing so or that any action would have been taken against them if so.

The importance of condition and provenance was well illustrated by the sale at auction in 1925–26 of the collection amassed by Sir Vauncey Harpur Crewe which, if reports were correct, was gathered at great expense and included a number of Hastings birds. Nicholson (1926) attended the sale:

> *The impression, even looking at the exhibition from the standpoint of a collector, was of an appalling and criminal waste. Here, dumped down anyhow in a series of dirty old cigar boxes and cardboard boot boxes, were enough eggs and enough dusty mounted skins to populate a county with birds. But rarity after rarity was treated by the bidders as rubbish simply because no data had been preserved to authenticate it, and even in the minority of cases where data were attached, they were far from unimpeachable. Some of the alleged identifications … were a standing joke all over the room.*

Patterson (1930) compared two sales of Norfolk collections that also illustrated the importance of provenance. From the sale of Alfred Masters's collection in 1886 he quoted prices for a sample of 12 lots containing, among other species, Goshawk *Accipiter gentilis*, Osprey *Pandion haliaetus*, Night Heron *Nycticorax nycticorax*, Cattle Egret *Bubulcus ibis*, Little Bittern *Ixobrychus minutus*, Golden Eagle *Aquila chrysaetos*, Little Bustard *Tetrax tetrax*, Cream-coloured Courser *Cursorius cursor* and Great Bustard *Otis tarda*. The 12 lots were sold for £8–19s (*c.* £900 today) and Patterson commented that 'the exceedingly low' prices must have arisen from lack of any proper details of where and how they were obtained. He contrasted this with 13 comparable lots from the sale of Henry Stevenson's collection in 1887, which were offered with exact details of the place where each bird was obtained, how it was taken, and the date. These lots made £103–1s–6d, something over £10,000 today.

Ruffs and godwits

It was not just what we would consider rarities today that were affected by this passion for unusual birds. I analysed the records of Black-tailed Godwit *Limosa limosa* in England and Wales up to 1914 after its disappearance as a breeding bird around 1835, and found that 70–75% of the 200 or so birds recorded were shot as specimens; the proportion in inland counties was often 100%. Waders such as Knot *Calidris canutus* and Spotted Redshank *Tringa erythropus* in summer plumage were much in demand (Patterson 1929) as were species such as Pied Flycatchers *Ficedula hypoleuca* outside their breeding range. Clearly not just national rarities were sought, but local ones as well.

Male Ruffs *Philomachus pugnax* in breeding plumage were particular favourites, because of the almost unlimited range of variation in their plumage. Plate 8 shows a case mounted by the Norwich taxidermist T. E. Gunn. Morris (2010) suggested that most of these birds were probably collected in the Netherlands and imported as cabinet skins. Like a number of leading taxidermists in his day, Gunn formed an important collection of British birds.

Another type of specimen eagerly sought was the unusual plumage variety. For birds, melanism, albinism, partial albinism and leucism were particularly dangerous genetic traits to inherit. Bolam (1912) described a pale cream-coloured (leucistic) form of Red Grouse *Lagopus lagopus* that occurred on the Blanchland Moors in Durham in the early 19th century, which Selby (1833) thought might have increased in number 'but for the anxiety of sportsmen to procure specimens'. Plumage varieties of the Goldfinch, known as *cheverels*, always commanded premium prices from collectors (and cagebird fanciers, by whom it was much prized for mule-breeding, where the finches were cross-bred with other species). The cheverel had 'the chin wholly white, with the white of the sides of the head extending upwards in a well-defined line through the black band and across the occiput; the brown patch on the side of the breast being also replaced by white' (Bolam 1912). Nicholson's (1926) description of the Vauncy Harpur Crewe collection noted that 'the quantities were amazing. Among the stuffed skins, gamebirds in endless variety and pied Blackbirds, Rooks *Corvus frugilegus* and Jackdaws *Corvus monedula* were the chief items', and one of the lots that sold well was a badly mounted but rare plumage variety of the Goldfinch.

Harrying rarities

One of the most active private collectors of the 19th century was E. T. Booth (Figure 10.3), who formed an extensive private museum at his house in Brighton and left a detailed account of his activities in his *Rough notes on the birds observed during twenty-five years' shooting and collecting in the British Islands*. A feature of his collecting was that he obtained all his specimens himself. His collection and museum were left to the Brighton Corporation on his death, who still maintain it. However, many collectors employed or commissioned gamekeepers and other rural workers to obtain specimens or eggs or information; to a shepherd, 2/6d

Figure 10.3. *E. T. Booth, bird-collector extraordinaire, and his museum. From Morris 2010, courtesy P. A. Morris.*

or 5/- (sums that are frequently mentioned in the avifaunas) for a rare bird or eggs would represent a significant proportion of the week's wage, which was worthwhile despite being very considerably less than the specimen's true value in the market. Rare birds and eggs thus came to have a clear monetary value to the people who were, perhaps, in the best position to regularly obtain them. Tubbs (1968) gave a good example of the malign impact of this kind of relationship in discussing the birds of the New Forest. He noted that when the Hon. Gerald Lascelles was appointed Deputy Surveyor of the Forest in 1880, he found 'that the keepers were all hand in glove with a collector or taxidermist. No bird whose skin was likely to make money was allowed any peace'. Rare birds of prey were particularly affected. It was a situation that Lascelles did much to remedy, but Charles St John (1848) suggested that similar relationships were common in the Scottish Highlands, as they undoubtedly were elsewhere.

All the 19th century avifaunas gave examples of the enthusiasm with which vagrants and other rare or unusual birds were harried. Nelson's (1907) account of the demise of the Great Bustard in Yorkshire is a case in point. I have long wondered whether this bird could have adapted to the rotation farming practiced in the early 19th century, but it was clear that it was never given the chance. Instead, the bustard was simply exterminated by people who never gave a thought to what they were doing. It was never necessary to shoot such birds to satisfactorily identify them. The story in East Anglia was much the same (Shrubb 2011). Many parallel cases could be cited. For example, St John (1848) apparently specialised in obtaining Ospreys for collectors. Finishing his account of destroying four breeding attempts to collect adults, eggs and young, he wrote 'there are very few in Britain at any time, their principal headquarters seeming to be in America; and though living in tolerable peace in the Highlands, they do not appear to increase nor to breed in any localities excepting where they find a situation similar to what I have described [an islet in a lonely loch]. As they in no way interfere with the sportsman or others, it is a great pity that they should ever be destroyed'. Crocodile tears with a vengeance.

The county avifaunas reported many similar cases. For example, Taylor *et al.* (1999) recorded that William Fisher, a naturalist (a term which at that time often meant a dealer in natural history artifacts) of Yarmouth, commissioned a man to obtain all the Goldcrests *Regulus regulus* he could in 1843, in the hope of getting a Firecrest *Regulus ignicapillus*, which he did, having shot 30 Goldcrests. In the great influx of Grey Phalaropes *Phalaropus fulicarius* on the south coast of England in autumn 1866, J. H Gurney estimated that some 500 were killed, around 250 of them along the Sussex coast (Yarrell & Saunders 1884).

Dealers habitually paid something for everything brought to them, to maintain the flow. T. Boyton procured 30 Little Gulls *Larus minutus* at Bridlington between 1868 and 1872, and another 30 were shot there between February 12th and 14th, 1870 (Nelson 1907). Nelson also recorded that, during an invasion of Red-necked Grebes *Podiceps grisegena* off the Yorkshire coast in January 1891, 28 were shot in the Scarborough area alone. Bolam (1912) recorded shooting 20 out of a flock of 50 Lapland Buntings *Calcarius lapponicus* at Ross, Northumberland, in early January 1893, most of which went into his collection. He also recorded that, in a great influx of Pomarine Skuas *Stercorarius pomarinus* in October 1879, 'flocks had appeared at the mouth of the Tweed, so fearless of human presence that the Spittal people had killed as many as they liked of them, one man having shot over a score to his own gun, and several other people nearly as many'. Large numbers appeared at the same time on the Berwickshire and East Lothian coasts, and Bolam (1912) reported that Mr R. Gray had 30 or 40 specimens submitted to him for examination. Gray also remarked on 'the recklessness' of the flocks which visited Dunbar, where they alighted in crowds on the pier and would not leave on being shot at.

The same man reported seeing numbers of Shovelers *Anas clypeata* shot on the Ribble in Lancashire early in May; he told Mitchell (1885) that 'in 1851 and 1852, along with the late Dr Nelson of Lytham, I often called upon a birdstuffer in Preston named Sharples, and it was on the occasion of these visits I had an opportunity of seeing the Shovelers and also Ruffs and Reeves in quantities. I have seen as many as 20 to 30 of each species in his hands at one time, all in the flesh' (Gray 1871). It is hardly surprising that the Shoveler had become rather a rare bird in Britain by the 1870s. Such activities created demand, but it is difficult to see any genuine scientific ornithological content or intent behind them.

It is easy to be condemnatory and dismissive of the activities of these collectors. But one point has to be made here. Indifferent optical equipment, at least by modern standards, and limited understanding of the identification characters of many scarce migrants meant that records of such birds depended to a large extent on specimens. Taxidermists were valuable sources of information to compilers of county avifaunas. Much basic ornithology and the understanding of the taxonomy of birds also depends on the availability of good series of specimens. So the taking of specimens was part of mainstream ornithology. But nothing justified the harassing of species such as Great Bustards to extinction, or shooting barrow-loads of Pomarine Skuas or Grey Phalaropes to satisfy the avidity of collectors.

Fraud

Collecting became intensely competitive, which added to the incentive for gunners to supply birds and which, allied with some worthwhile sums of money, led to fraud. The most notable example was the Hastings Rarities (Nicholson & Ferguson-Lees 1962; see also, for example, Greenwood 2012, Harrop *et al.* 2012), but Melling (2005) drew attention to another series, from Yorkshire, which he styled the Tadcaster rarities. These had a similar history. I would be very surprised if other series do not lurk undetected in collections and the record books.

The authors of the county avifaunas were certainly aware of such problems, however. Ticehurst (1909), for example, discussed a series of records obtained by Stephen Mummery, a taxidermist and collector operating in the Margate area in the early 1840s. He was responsible for a large number of specimens in the Margate Museum, which was opened in 1839 and closed, and its collection sold, in 1868. Mummery's records between 1840 and 1844 included a distinguished group of rarities, embracing Cream-coloured Courser *Cursorius cursor*, Great Snipe *Gallinago media*, Spotted Sandpiper *Actitis macularia*, Richard's Pipit *Anthus richardi* (suspected of having bred), Woodchat Shrike *Lanius senator*, Nutcracker *Nucifraga caryocatactes* and Rose-coloured Starling *Sturnus roseus*, as well as a group of birds rare then but now established as uncommon but regular migrants. Although he included them in his book, Ticehurst was politely sceptical of records from this source.

Ticehurst also discussed a group of records claimed by James Green, a London dealer and taxidermist, for north Kent in the early 1850s. These comprised a Bluethroat *Luscinia svecica* of the white-spotted race *cyanecula*, two Great Reed Warblers *Acrocephalus arundinaceus,* and the nest and eggs of Savi's Warbler *Locustella luscinioides*. The records were all rejected, because Green was known to import such species alive and as specimens from the Netherlands.

These incidents illustrated well the problems surrounding records resting on specimens, and Ticehurst's account of the early records of Pine Grosbeak *Pinicola enucleator* in Kent outlined similar problems. Kelsall & Munn (1905) reported four Pine Grosbeaks said to have been obtained in Woolmer Forest, Hampshire but stuffed in Great Yarmouth, which were suspected to have been imported on ice from Russia. O. V. Aplin reported receiving two from Russia at Leadenhall market at the same time.

Coward (1922) went further, writing that

> *greedy collectors, by no means an extinct class, have only themselves to thank for the fraud which surrounds the 'identity' of species. A school of collectors and taxidermists discovered that there was a market for British or locally obtained specimens which could be supplied by substituting specimens from abroad or from other areas than those stated on the labels. Wildfowl and often ornithological oddments arrived from abroad in the wholesale markets and these were speedily snapped up, and often shown to the collector 'in the flesh' with an entirely spurious local history.*

Coward remarked that the skins of Ruffs were imported from Holland, mounted and sold as locally obtained, and that American skins were treated in the same way to obtain big prices for British-killed rarities.

Imports on ice

There is no doubt that large numbers of suitable specimens were being imported into Britain, dead and alive, by the late 19th century and were readily available in markets in London, Liverpool and Manchester. Collectors regularly trawled market stalls and game-dealers' stock in search of specimens to add to their collections. But the difficulty with all such records was certainty of source. Gurney (1870) wrote that 'I am quite certain that a large proportion of the birds in the London markets are British-killed, and the game dealers, when they get to know you, will give not only the exact locality, but occasionally even the names of their agents', which might not strike the modern eye as an entirely satisfactory basis for records of rare birds. Gurney (1883) was later told by Brazenor, a taxidermist in Brighton, that 100 Waxwings had been sent with one consignment of game from Russia. Gurney remarked that 'these were no doubt intended for the benefit of bird-stuffers', to whom they were, in fact, sold; Brazenor had several mounted in his shop. Stubbs (1913) recorded that 'I have seen Waxwings, Pine Grosbeaks, Scarlet [*i.e.* Northern] Bullfinches, Crossbills [*Loxia curvirostra*] and other northern forms in great quantities in the Manchester market'. He also recorded finding numerous specimens of Baikal Teal *Anas formosa* and Falcated Duck *Anas falcata* in Leadenhall Market in a consignment of wildfowl from China.

D'Urban & Mathew (1895) reported a particularly interesting example of the importation of frozen King Eiders *Somateria spectabilis* in the late 19th century, which were purchased by collectors from Plymouth market, although there was no evidence here of any attempt to palm them off as British-taken. In the Brighton area, and no doubt elsewhere, ornithologists regularly checked the bird-trappers' cages in the field for unusual birds. Booth (1881–87) commented that 'during severe weather in Sussex I often remarked [noticed] a fresh-captured Shorelark [*Eremophila alpestris*] or two in the store-cages of the bird-catchers'. Walpole-Bond's (1938) accounts also made it clear that many of the rarer birds he listed from the Brighton area were seen alive in the dealers' cages.

Frost (1987) remarked that certain species were always more popular with the occasional collector, who merely wished for a case or two for decoration. Bitterns *Botaurus stellaris*, owls, Green Woodpeckers *Picus viridis* and Kingfishers *Alcedo atthis* were particularly vulnerable to this demand, and the last three also for fire screens, shallow glass-fronted panels filled with an assortment of colourful birds, native and imported, used to decorate the fireplace when not in use (Morris 2010). M. A. Mathew noted one bird-stuffer in Taunton who received more than 50 Green Woodpeckers for preservation in the weeks after Christmas 1877, and Hartert & Jourdain (1920) recorded a bird-stuffer in Great Marlow handling around 100 Kingfishers in 1890.

Stevenson (1870) cited the Bittern as a bird constantly shot in the mistaken belief that it was a rarity. He found that 108 specimens had come to his notice in Norfolk in 18 years, mainly from the Broads, from where up to 20 were brought into Yarmouth annually. Most were winter visitors, and often sold for no more than 1/6d to 2/6d. They were also regularly taken in north Norfolk, where they were sometimes extremely plentiful in the King's Lynn area. It is small wonder that Stevenson regarded the activities of gunners and collectors as mainly responsible for preventing Bitterns recolonising Norfolk. Nor was this pattern unique to Norfolk. McAldowie (1893) noted that stuffed Bitterns, obtained in the district, were to be found in many of the cottages around Eccleshall in Staffordshire, and Bull (1888) found that many specimens existed in houses throughout Herefordshire. In a sample of ten other 19th-century county avifaunas, I found that 172 of 203 records of Bitterns were of birds that had been shot. This was clearly a tolerably common bird, and the demand came from those who wanted a fashionable decorative object, rather than the dedicated collector. Bramblings were particularly popular for this purpose in Cheshire, and Coward & Oldham (1900) noted that the cases of stuffed birds so common in cottages and farmhouses there nearly always contained one or two. But the Barn Owl *Tyto alba* was probably the species most frequently found stuffed, often badly, in such places. Mathew (1877) recorded two bird-stuffers in Taunton who each received an average of 50 Barn Owls and 40 Tawny Owls *Strix aluco* annually for preserving; another bird-stuffer in the town did a similarly large trade in mounting owls. They were frequently made into hand screens 'stuck upon handles with the head, legs and feet in impossible positions between the wings, and with staring glass eyes, almost always of the wrong colour' (Lilford 1895). Short-eared Owls *Asio flammus* were similarly treated (Boyes 1877).

Egg-collecting

Although Chalmers-Hunt (1976) recorded the auction of an egg collection as early as 1789, egg-collecting became a major feature of collecting activity from around the mid-19th century (Figure 10.2, p. 179). Specialist dealers often operated by post, issuing regular catalogues of recent acquisitions (Cole 2006). Cole concluded that, with a few important exceptions, the dealing industry for eggs never seemed to have progressed from a 'hotch-potch of bit-players' serving the lower end of the market, because most activity involved dealing and exchanging between collectors or through auctions, where provenance could better be assured. The chief auction house was Stevens Auction Rooms in Covent Garden, London, which held important sales of eggs from 1850 to 1939. Cole noted that this place provided a 'perfect clearing house' for the best material available, and had the best information.

Effect on populations

Egg-collecting undoubtedly caused significant damage to the populations of some scarcer breeding species. For example, Stevenson (1866) and Ticehurst (1932) both considered that egg-collecting, aggravated by shooting specimens, was a major factor in the disappearance of Bearded Tits *Panurus biarmicus* from many of its haunts in East Anglia. Gurney (1899) noted that there had been a systematic trade in their eggs for years, the recognised price of which was 4d apiece (about £1.70 today). He quoted H. A. Macpherson that Norfolk eggs were exported to Europe; Smith of Yarmouth, who, for example, handled 113 eggs between April 10th and 20th 1876 (Riviere 1930), also supplied a great number to other dealers in Britain, and from that source many collections were supplied. Gurney estimated that the total population of the Broads declined from 160 nests in 1848 to 33 in 1898, and recorded that, by 1889, only two pairs remained at Hickling and Heigham, which had once been the species' stronghold. The bird persisted under protection on small private broads, and statutory protection from 1895 started to benefit the species. Elsewhere in East Anglia, the Bearded Tit was either almost or totally extinct by the end of the 19th century. Drainage, of course, played a major part in its demise in East Anglia and in other parts of southern and eastern England, where it had disappeared even earlier.

The Dartford Warbler *Sylvia undata* suffered a similar fate at the hands of collectors. First discovered at Dartford, Kent in 1773 and on Wandsworth Common 10 years later, Bucknill (1900) noted that 'there seems no doubt that the excitement occasioned by the occurrence of this bird and the anxiety of collectors to obtain specimens very quickly, almost exterminated the species in the neighbourhood of London, where in the early part of the [19th] century it would appear to have been not very uncommon'. It remained locally distributed elsewhere in Surrey. In 1886, Booth (1881–87) wrote

> *A few years ago Dartford Warblers were to be found in many of the large patches of furze scattered over the South Downs; but the constant demand for their nests and eggs by collectors and dealers has, as might have been expected, at length thinned down their numbers considerably, and many of their former haunts are now deserted.*

Walpole-Bond (1938), however, believed that severe winters were a more important factor in the bird's demise. Either way, unbridled collecting and bad weather would have made a lethal combination. Walpole-Bond himself was a notable egg-collector, and the Dartford Warbler was one of his favourite species. Reading his Sussex diaries leaves one with the strong impression that he was also largely responsible for the disappearance of the Marsh Warbler *Acrocephalus palustris* as a Sussex breeding bird and, indeed, that he was collecting their eggs commercially.

Ticehurst (1909) said much about the damage inflicted by egg-collectors on Stone Curlews *Burhinus oedicnemus* and Kentish Plovers *Charadrius alexandrinus* at Dungeness in Kent. A systematic trade was carried on in the eggs of Stone Curlews, each pair having a traditional territory used annually and well-known to the local fisher lads. At least one local inhabitant acted as an agent, buying every clutch he could at four to five shillings per clutch (£20–£25 today) to sell on in London. Common on Dungeness in 1844, the species had largely disappeared by 1890. Protection afforded by the creation of the Army firing range and the appointment of a Watcher by the RSPB from the end of the 19th century enabled some recovery. The Kentish Plover, which always had a very limited breeding range in Britain that was largely confined to beaches and shingle areas of southeast England, was similarly harassed. Yarrell (1845) learned from collectors in the Hastings area in 1833 that dogs were trained to hunt for nests and eggs, over extensive tracts of breeding territory along the shores of Kent and East Sussex. 'On finding a nest of eggs,' he wrote, 'which they did by scent, the parent birds being in some instances upon the nest, the dog stopped till the master came up to examine the ground, and this done, the dog went off again, upon signal, pointer-like, to hunt as before'. Such activity, combined with persistent shooting of breeding pairs, had, by the end of the 19th century, virtually exterminated a species first recorded in 1787, which early accounts suggest was once quite numerous within its restricted range.

These are just a few examples of the rarer species that egg-collecting, often combined with habitat loss or persecution, helped to reduce or exterminate. There are many others, especially among the raptors. The Red Kite *Milvus milvus* was a particular example. Exterminated in England by game-preserving interests, it hung on in the Welsh Hills on the borders of Carmarthenshire, Breconshire and Ceredigion, where it was much harassed by egg-collectors. In 1903, Dr J. H. Salter persuaded the British Ornithologists' Club to start an effective nest-protection scheme, which saved the kite from extinction. It rested on a system of payments to landowners to protect nest-sites that lasted almost to the end of the 20th century, by which time the Welsh Red Kite population had finally achieved a sound basis of some 400–600 territorial pairs (Welsh Bird Report 2009; Roderick & Davis 2010). Peregrines *Falco peregrinus*, Ravens *Corvus corax*, Ospreys and White-tailed Eagles *Haliaeetus albicilla* were similarly harassed, the last two to extinction in the early 20th century.

Species targeted

Collectors often concentrated on particular species. The profiles given by Cole & Trobe (2000 & 2011) show that species such as Eurasian Hobby *Falco subbuteo*, Red-backed Shrike *Lanius collurio*, Cirl Buntings *Emberiza cirlus*, Dartford Warbler, Grasshopper Warbler *Locustella naevia*, Raven, Peregrine, Dotterel *Charadrius morinellus* and other waders were particular favourites. Such species recur again and again in Cole & Trobe's accounts. Another popular approach was the collection of long series of clutches of the same species, to show variations in colour, markings and size. Erythristic (abnormal red-marked) eggs, particularly of Ravens and Lapwings, were much sought after. Such series were not necessarily of rare species; Chaffinches, Blackbirds and House Sparrows *Passer domesticus* often figured, for example, because of the great variety in their eggs. Cuckoos *Cuculus canorus* with their hosts were especially popular and many collectors built important series.

But egg-collecting had a much wider basis. Old catalogues and auction sales suggested that many collectors probably aimed for a representative collection. They often made collecting trips abroad, particularly to Scandinavia and Spain. A list issued by C. H. Gowland in 1947 (in Cole 2006) showed that he included in his British List eggs of a large number of species that did not breed in Britain but had occurred as migrants or vagrants. This was just one aspect of a significant international traffic. Lilford (1895) reported that the eggs of Black Terns *Chlidonias niger* and many other marsh birds breeding in the Netherlands were commonly available in Leadenhall Market in his day. Gowland, who was probably the most important egg dealer in Britain of the first half of the 20th century, advertised that 'I have the largest stock of birds' eggs in Europe for disposal. Own private collectors in Iceland, Greenland, Scandinavia, Spain and other principal breeding localities in Europe and North Africa. Large supplies received annually from my own collector in India' (Cole 2006).

The guarantee of genuineness he gave was both valuable and necessary. It was so easy to make up clutches. J. E. Griffith told the story of visiting Lake Mývatn in Iceland and the way the eggs of the wildfowl that bred there were farmed. The nesting islands in the lake were divided between families of local farmers; the nests were visited daily and the eggs taken, leaving two to encourage continued laying. The eggs were sold to a dealer, W. F. Paulsson (in Griffith 2000), who

> *… then passes them off as genuine to collectors all over the world. The eggs arrive by the bucketful every evening, all unmarked; as many as four buckets each from maybe half-a-dozen farmers. The farmers wait in line and as each egg is removed from a bucket, Paulsson candles it and places it in line on a large dining table under species. When the last farmer had finished, all incubated eggs having been returned, he opens his lists. "I have a collector who wants a c/7 [clutch of seven] Whooper Swan Cygnus cygnus". He examines a line of perhaps 20 Whooper eggs and chooses seven that are somewhat the same size and shape! "That will do him" and so on,' all the time drinking whiskey straight from the bottle.*

G. H. Lings (in Cole 2006) told a rather similar story about Gyrfalcons *Falco rusticolus* and concluded that he was sceptical thereafter about clutches bought from Iceland 'and, indeed, anywhere else'. Knox (1850) made a rather similar point when discussing the taking of Golden Eagle eggs, which he suggested were worth £1 to 30/- each (£100–£150 today). He noted that

> *… indeed I have known a larger sum given for a very ambiguous looking specimen in England, 'warranted from the golden eagle' but which to an experienced eye had an unmistakeable look of having emanated from a Norfolk turkey yard.*

Some collectors built up international networks of agents and collectors to supply them with eggs from all over the world. One such was Vivian Hewitt, who lived at Cemlyn Bay on the north coast of Anglesey from the early 1930s, buying the site and creating the nature reserve now owned by the National Trust. In the 1920s and 1930s he actively collected eggs on his own account, often operating from a boat around the north coasts of Wales and Anglesey and specialising on seabirds, raptors and Choughs *Pyrrhocorax pyrrhocorax*. He gave up collecting personally as he became more interested in conservation, but continued to purchase existing collections from all over the world; his biographer, William Hywel, claimed that that was all he did, but his account of Hewitt's activities leaves the clear impression that Hewitt was still employing agents to collect for him in places as diverse as India, the Dutch East Indies (Indonesia) and the Falkland Islands (Hywel 1973), although this impression may perhaps be the result of poor editing.

Hewitt assembled a vast egg collection, which was bequeathed to the British Trust for Ornithology after his death in 1965. It was delivered to the BTO in four removal vans, and its then-director, Jim Flegg, estimated that the collection contained around 500,000 eggs. From the account he gave to Hywel, it is clear that the collection had overwhelmed Hewitt and he had done little to reduce it to systematic order. Much data was missing, and its scientific value can only have been limited.

Egging in summary

As with the collectors of stuffed birds, it is easy to be dismissive of the activities of egg-collectors in the past. But classic works of British ornithology, such as Witherby *et al.*'s *Handbook of British Birds* or Bannerman's *Birds of the British Isles*, show that egg-collectors made a major contribution to our knowledge of birds' breeding biology, though the same cannot be said of egg-collectors today.

One fortunate result of this passion for collecting eggs arose in the second half of the 20th century. For it was the existence in collections of large numbers of Peregrine clutches, and also those of Sparrowhawks *Accipiter nisus*, Merlins *Falco columbarius* and Golden Eagles, that enabled Derek Ratcliffe and his colleagues to work out the effect that the organochlorine compounds being used in agriculture were having on the breeding success of these species, by thinning their eggs' shells to the point of unviability. This work led to the banning of these substances, and the gradual recovery of the species concerned.

It is interesting to consider the motivation behind egg-collecting. Cole & Trobe (2000) quoted two perceptive observations by collectors on this. The first came from Desmond Nethersole-Thompson, who was adamant that in order to study breeding birds, one had to have a 'predatory hunger for the nest'. This observation had a somewhat wry resonance since, in his early days in Scotland, Nethersole-Thompson supported himself and his family by collecting and selling eggs, remarking that 'to eat we had to find …we usually ate'. The second observation was made by the Rev. J. R. Hale from Kent, who said that what appealed to him was that collecting eggs by oneself was the finest of all field sports, requiring more brains than any other, and absolute physical fitness. As one who spent much of his youth in pursuit of birds with a gun, I find that very convincing. It is worth remarking, in support of Hale's contention, that many of the favourite species with collectors were those whose nests required much skill and knowledge to find, or great effort to reach and take. It is of interest that Cole & Trobe's (2000 & 2011) profiles include remarkably few clergymen (only six), when one considers how many of the Victorian collectors of birds were parsons. Egg-collectors came from every walk of life, but 73% of those for whom their profession or job was recorded were professionals – doctors, solicitors, businessmen, serving officers in the armed forces and so forth, together with those of independent means and a handful of the clergy. The largest categories were serving officers in the forces, who often pursued some other vocation after retirement, doctors and businessmen. There is probably little to be read into this brief analysis, except that such professions probably provided the time for the extensive pursuit of collecting activities.

Egg-collecting was far from finished in Britain by the 1954 Protection of Birds Act, which prohibited the sale or exchange of eggs. Cole & Trobe (2000) show that many of the collectors they discussed were active in Britain in the 1960s and 1970s, and even into the 1980s. Unfortunately collecting still continues, albeit as a furtive and illegal activity.

CHAPTER 11
The plumage trade

Before discussing this subject, the term 'plumage trade' needs definition. Birds' plumage has been used for centuries, although many of the uses discussed earlier in this book, in Chapter 5 or 8, for example, tended to be utilitarian. Nevertheless, using the feathers and skins of birds, particularly brightly coloured species such as birds-of-paradise, hummingbirds or kingfishers, or those with elegant and beautiful plumes such as egrets, as adornments also has a long history. Until the 19th century, however, the impact of such exploitation as far as Europeans were concerned tended to be limited by cost, a lack of knowledge of many of the most desirable species' distribution, and sometimes the privilege of rank.

Ostrich plumes were perhaps the earliest manifestation of the use of such feathers for ornament. The Ostrich *Struthio camelus* was valued for its plumage by the Greeks and Romans in classical times; hunting scenes in Roman villas included Ostriches as game animals, and showed dignitaries wearing their feathers as ornaments. With the fall of the Roman Empire, the use of Ostrich plumes largely disappeared in Europe until the late mediaeval period; they began to appear regularly on headwear in 14th-century France, and were highly prized and very expensive, particularly white feathers from the wings of the male. They were taken from birds hunted in North Africa and came into Europe through Italy, particularly Venice. Ostrich plumes became an important component of the extravagant head-dressings fashionable in Europe, particularly the French court, in the later 18th century (Doughty 1975).

Doughty observed that Ostrich-farming started in the late 19th century in Cape Colony (part of modern South Africa) and North Africa, and spread to Australia, Europe and the United States. Chevenix Trench (1922) remarked that such farms, mainly in South Africa, were the principal source of Ostrich plumes to the London market, and suggested that such farming had been instrumental in preventing the Ostrich's natural distribution from becoming confined to 'the least accessible parts of Africa and Asia'. After 1921 farmed Ostrich feathers (and goose feathers) provided the main components of the London feather trade, and Penry-Jones (1958) gave a detailed description of the management of the trade through the Port of London Authority's warehouse in Cutler Street.

By the 19th and early 20th century the plumage trade had become a matter of fashion, and exploiting such adornments became largely a question of money. Three factors probably combined to encourage the great expansion of the trade, particularly from the middle of the 19th century. The first was an increasingly accurate understanding of bird distributions and numbers, and where the most desirable species were to be found. The second was a rise in interest in fashion, allied to increasing personal wealth, particularly in the growing middle classes. The third was the greatly increased reach, capacity and reliability of transport, as important in this trade as in other facets of the exploitation of birds.

Whilst plundering bird populations for their plumage was a global activity, the main centres or clearing houses for this trade were Paris, London and New York; Paris was the main centre for processing and finishing feather millinery, with 10,000 people employed in this work in the city and surrounding areas in the late 19th century (Doughty 1975, del Hoyo *et al.* 1992). A significant part of the trade in both London and New York comprised imports from Paris. Berlin and Vienna were also important centres until the 1914–18 War destroyed the trade there (del Hoyo *et al.* 1992). Thus the trade had an important European focus. A map published by Doughty (1975) suggested a marked tendency for overseas sources of birds to London and Paris to be areas where Britain and France had existing colonial interests.

Egrets and egret-farming

An esoteric aspect of this trade was the development of egret farms in parts of the Indian subcontinent, to provide plumes for the London and domestic markets. Such farms were mainly concentrated in the province of Sind, and were managed particularly by the Mirbahars (the peoples of the inland waters of the region; Birch 1921), who mainly lived by fishing and catching wildfowl, and already kept herons for use as decoys in netting wildfowl (see p. 57). In the later 19th and early 20th centuries, egret farming had largely replaced slaughtering egrets as a source of plumes in Sind (Birch 1921).

The birds were kept in units of 80–120 birds in enclosures or runs, roofed and walled with matting or reeds and measuring about 40 x 20 x 5 feet (12 x 6 x 1.5m), supplied with abundant clean fresh water and fed on small fish or fry (Benson 1922). Those reporting this enterprise were unanimous that the birds, being valuable, were kept in good conditions and were well looked after. Birds were pinioned, and Birch (1921) noted that they bred twice annually and moulted four times, twice in summer and twice in winter; the two summer moults gave only a light return. Birch commented that this was regarded as impossible in London, but noted that he had reliable evidence that it was correct. Birds were plucked of their plumes while alive, a process that was reported to apparently do little harm, although Chevenix Trench (1922) considered that methods needed improvement.

Although the original base population of captive colonies stemmed from birds trapped in the wild, by the time Birch (1921) wrote they were wholly maintained by captive breeding, with the tame birds nesting on the ground, using nest materials supplied by their keepers. Birch knew of at least 100 egret farms in the province, each containing some 100 birds, and he recorded that some farms had been started in the Punjab and Assam. Doughty (1975) also recorded that egret-farming was being practiced at this time in southeast Asia, particularly in Burma, and that the French were experimenting with it Tunisia in 1895.

The *Journal of the Bombay Natural History Society* discussed this rural industry in some detail, urging that the legislation prohibiting the sale of egret plumes was ineffective, and widely circumvented by smuggling. Chevenix Trench (1922) observed that 'their plumes are so valuable and so easily smuggled out of the country, that no amount of prohibitive but unintelligent legislation will prevent the export trade in *aigrettes*, which continues today as it has done in the past, in spite of the Plumage Acts'. And 'in November the Central Provinces are overrun by iterant middle men, who buy up plumes from petty local men and either take them to a 'gentleman' in Calcutta or hawk them about Indian cities for a price which, in my experience, ranges from 10 to 28 times their weight in silver'. Chevenix Trench supported his arguments by pointing out that Ostrich farming had been largely instrumental in preventing a serious diminution in the Ostrich's natural distribution. He argued that to save the egrets in India, egret-farming should be encouraged and the export of farmed plumes permitted under licence. Correspondents were unanimous that a valuable rural enterprise could be created that would help meet the demand for plumes and benefit wild populations of egrets, which were being damaged by smuggling but which were unharmed by egret-farming as it was managed. The matter was never resolved.

Domestic sources

While Paris sourced many birds from within France, there were few such domestic sources available to the fashion trade in London. Nevertheless, some species were heavily exploited in England for the trade. One such was the Great Crested Grebe *Podiceps cristatus*, a species that was locally numerous in England before the drainage of major wetlands in the 18th century (Brown & Grice 2005). Pennant (1776) noted that grebe skins, which were prized for their beauty and softness and known as grebe-fur, were imported into England from Lake Geneva in the 18th century, selling for 14/- each, a price also quoted by Bewick (1826). Some were also taken from the meres of Shropshire and Cheshire and from the Fens, selling for the same price, but it was only a minor trade; it is of interest that Ray (1678) appears not to refer to it.

Harrison & Hollom (1931) noted that attitudes to the species changed with the publication of a letter in the *Zoologist* of 1857 by a Mr Strangeways, which reported that

> *In the months of April and May last I collected 29 of these birds in full summer plumage, all shot in Norfolk. Three of them are preserved and they are now in the Great Exhibition in Hyde Park, where they are exhibited by Messrs. Robert Clarke & Sons, the furriers, in Class 18, to which they very appropriately belong, as the breast of this bird has become a fashionable and very beautiful substitute for furs ... the rest of the skins I have manufactured into ladies' boas and muffs.*

Strangeways commented that the market for grebe-fur at that time was mainly supplied from southern Europe. Harrison & Hollom (1931) considered this letter to be the direct cause of the severe persecution of the species that subsequently occurred in England, particularly in East Anglia. As early as 1860, the English population had been reduced to about 42 pairs, and it would probably have become extinct but for the existence of pairs on protected waters. The species recovered rapidly with the passing of the Birds Protection Acts in the late 19th century, and Harrison & Hollom noted that 45 new waters were colonised between 1880 and 1890; this seems to have been a part of an expansion in range in Norway and Sweden, so protection may not have been the only factor involved in the grebe's recovery.

Seabirds

Seabirds were also exploited for the plumage trade in Britain, particularly Kittiwakes *Rissa tridactyla*. Yarrell & Saunders (1884) gave a detailed account of the trade in Kittiwake wings at Lundy. They recorded that many young were still in the nest or barely flying when the close season set by the Seabirds Protection Act of 1869 expired (on 1st August), and the complete wings were in great demand for decorating ladies' hats.

> *Vast numbers were slaughtered at their breeding haunts. At Clovelly, opposite Lundy Island, there was a regular staff for preparing the plumes, and fishing smacks with extra boats and crews used to commence their work of destruction at daybreak on the 1st of August, continuing this proceeding for upwards of the fortnight. In many cases the wings were torn off the wounded birds before they were dead, the mangled victims being tossed back into the water; and the Editor has seen hundreds of young birds dead or dying of starvation in the nests, through want of their parents' care, for in the heat of the fusillade no distinction was made between old and young.*

Yarrell & Saunders estimated that at least 9,000 birds were destroyed during the fortnight. Other authors reported this trade in Kittiwakes. Nelson (1907) noted that one gunner operating at Flamborough cliffs claimed to have made up to £18 per week (£1,400 today) during the season and that, in one season, 4,000 birds passed through his hands *en route* to the London *plumassiers*. The birds were shot when building their nests, at which time they were easily taken as they collected nest material from the cliff tops and adjacent fields.

Mathew (1869) also recorded large-scale slaughter of Kittiwakes in winter at Weston-super-Mare in Somerset. The birds followed shoals of sprats into the shallow waters of Weston Bay where 'our fishermen found that, in consequence of the great demand for gulls' wings for ladies' hats, it would be quite as profitable for them to shoot the birds as to attend to their nets'. The birds were attracted by throwing broken sprats overboard, and the men could bring back 40 or 50 each after a morning's work. Mathew remarked that he had more than once met men returning with large panniers full of dead gulls.

Terns were widely shot on breeding colonies in East Anglia and probably elsewhere on the east and south coasts of England. Shooting took place, as with Kittiwakes, after the end of the close season at the beginning of August, usually by gunners from nearby towns (Christy 1890, Stevenson & Southwell 1890; these authors stated that the birds shot were simply left to rot, though there must have been demand from the fashion trade).

Range of species taken

A wide range of species was involved in the international trade, although the publicity given to its barbarities in the late 19th and early 20th centuries concentrated on a few, especially egrets and other herons, and terns, largely because of the way they were harvested. But Scott (1887), discussing the destruction of heronries in Florida, made the point clear, reporting of May 25th 1886 that

> *In the morning I went on the beach with Mr Batty, and we shot Knots [Calidris canutus], Black-bellied Sandpipers [Grey Plover Pluvialis squatarola], Sanderlings [Calidris alba] and Turnstones [Arenaria interpres] over decoys, all these species being used by Mr Batty in his feather business. At the same time two of Mr. Batty's men were killing Wilson's Plovers [Charadrius wilsonia], Least Terns [Sterna antillarum], Boat-tailed Blackbirds [Quiscalus major], Gray Kingbirds [Tyrannus dominicensis] and any other small species that came their way. The Least Terns are particularly in demand in the hat business, and Mr. Batty paid for such birds as I have enumerated ten or fifteen cents each in the flesh. All owls, and particularly the Barred Owl [Strix varia] are desirable. The feathers of these, as well as of hawks, are bleached by processes that Mr. Batty described to me, and used for hats and other decorations. One of Mr. Batty's employees told me that they had left a party at the pass below, where they were killing the same kinds of birds, and that Mr. Batty was constantly purchasing and trading with native and other gunners for plumes and round and flat skins of all the desirable birds of the region. Not less than sixty men were working on the Gulf coast for Mr. Batty in this way.*

At the other end of the trade, Table 11.1 summarises the range of bird species used in the millinery trade in New York, as reported in *Harper's Bazar* (sic.) from 1875–1900. Considerably more species than the 27 noted here were involved, as these entries comprised species groups. But they give good support to Scott's (1887) observations, and it is worth noting that Doughty (1975) recorded that 83,000 people were employed in the making and decorating of hats in the United States at this time, mainly in New York.

Species	Species	Species
Hummingbirds	Orioles	Herons
Pigeons	Grebes	Partridges
Larks	Pheasants	Domestic fowls
Blackbirds	Doves	Swallows
Starlings	Ostrich	Owls' heads
Birds-of-Paradise	Scarlet Ibis	Parrots
Marabou	Kingfishers	Ducks
Peafowl	Paroquets	Seabirds
Tanagers	Impeyan Pheasant	Guineafowl

Table 11.1. *Birds or groups of birds used in millinery in New York, as reported in Harper's Bazar (sic.) from 1875 to 1900. Source Doughty 1975.*

A snapshot of the scale of the trade through London was provided by a note in the *Auk* of July 1888 (pages 334–35), which summarised the catalogue of an auction sale held at Hale & Sons of Mincing Lane, London on March 21st, 1888. The author noted that advertised for sale were

> *birds' skins, plumes, wings and feathers, representing in the aggregate more birds than are contained in all the ornithological collections of this country, including private collections as well as public museums – in other words, hundreds of thousands, in this single auction sale! Besides about 16,000 packages and bundles of 'Osprey', Peacock, Argus and other Pheasants, Ducks, 'Paddy' and Heron feathers, we note several thousand mats and hand screens, while under the heading of 'various bird skins' we figure up between 7000 and 8000 Parrots, shipped mainly from Bombay and Calcutta, but including some from South America; about 1000 Impeyan and 500 Argus Pheasants; about 1000 Woodpeckers; 1450 'Penguins' (Auks and Grebes?); some 14,000 Quails, Grouse and Partridges; about 4000 Snipes and Plovers; about 7000 Starlings, Jays and Magpies; over 12,000 Hummingbirds; about 5000 Tanagers; 6000 Blue Creepers and 1500 other Creepers (probably honeycreepers); several hundred each of Hawks, Owls, Gulls, Terns, Ducks, Ibises, Finches, Orioles, Larks, Toucans, Birds of Paradise etc; 1493 Swallows in one lot; and about 12,000 are scheduled under 'Black Heads', 'Black and White', 'Pink and Black', 'Grey and Black', 'various' etc. The number distinctly scheduled as skins reached nearly or quite 100,000, while the number represented by the 16,000 or more packages and bundles and the 3500 mats and hand screens must amount to at least as many more.*

The author remarked that such sales were not infrequent and the traffic, if sustained, must have markedly affected populations of the birds involved in the countries supplying the sales. He went on to report that, in a recent edition of the *American Field*, it was stated that

> *… last year the trade in birds for womens' hats was so enormous that a single London dealer admitted that he had sold 2,000,000 small birds of every kind and colour. At one auction in one week were sold 6000 Birds of Paradise, 5000 Impeyan Pheasants, 400,000 Hummingbirds, and other birds from North and South America, and 360,000 feathered skins from India.*

Other statistics quoted by del Hoyo *et al.* (1992) recorded that 750,000 bird skins were sold in London in the first quarter of 1885 and that one Florida merchant sent 130,000 skins to New York in 1892.

Doughty (1975) observed that no exhaustive list of birds used in the trade had ever been compiled and that there was a paucity of trade figures available. Nonetheless, he compiled some statistics extracted from London auction sales from 1890–1911, which add further evidence to those above. These showed, for example, that 39 sales of herons and egrets between 1897 and 1911 had involved 1,215,000 birds, that 51,200 Sooty Terns *Sterna fuscata* had been sold in three sales in 1908, 50,000 kingfishers in three sales in 1906–1907, and 152,000 hummingbirds in 8 sales during 1904–1911.

Figure 11.1 summarises the total weight (in the UK) and total value of imports of feathers into the United Kingdom and the United States between 1872 and 1930, by decade. The figures show a steady increase during the period, interrupted in Britain by the collapse of the trade in Europe caused by the 1914–18 War. At its peak, the total value of the trade in the UK alone was some £20 million per year (£1.82 billion today). In both countries, the trade

(a)

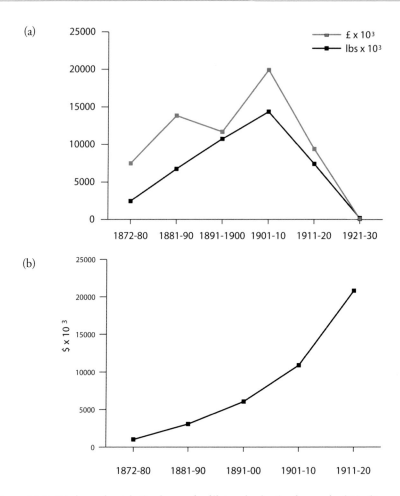

Figure 11.1. *(a) the total weight (in thousands of lbs) and value (in thousands of £s) of imports of feathers into the United Kingdom by decade between 1872 and 1930; (b) the total value (in thousands of $s) of imports of feathers into the United States by decade between 1872 and 1920. Data from Doughty 1975.*

declined sharply after 1920, with increasing legislation against it and with changes in fashion and attitudes resulting directly from the war. The total weight of feathers indicated as passing through the London market in the period 1872–1930 amounted to just under 19,000 metric tonnes. But this was less than half the 50,000 tonnes estimated to enter France in the same period. Both figures represent scores of millions of birds (Doughty 1975).

Worldwide carnage

The destruction the plumage trade inflicted on bird populations was appalling. Perhaps the best documented species are the herons and some of the seabirds, particularly terns. Voisin (1991) noted that three species of heron, Squacco Heron *Ardeola ralloides*, Little Egret *Egretta garzetta* and Great White Egret *Ardea alba*, were virtually exterminated in Europe in the late

19th and early 20th centuries by plume-hunters. But it should be noted that the Little Egret, at least, had already disappeared from much of northwest Europe in the mediaeval and early modern periods (see p. 58). The killing of herons was carried out in their colonies in the breeding season, when the plumes were at their best, and resulted not only in the death of the adults but also the loss of eggs and young, which were predated or starved.

It was a similar story in the United States. Scott (1887) described a trip down the west coast of Florida in May and June 1886, to revisit heronries he had visited six years earlier. He provided a graphic description of the havoc inflicted in that short period on the numerous colonies he had known. Altogether he revisited at least 30 large colonies along the coast from Tarpon Springs south to Punta Rassa, all but one of which had been shot out and was deserted; the one still occupied had 'no more than a dozen pairs of herons where thousands had bred six years before'. Colonies were of mixed species and the largest, which contained many thousands of pairs of herons of eight species plus Brown Pelican *Pelecanus occidentalis* and Double Crested Cormorant *Phalacrocorax auritus*, took just five breeding seasons to destroy. Scott noted piles of dead herons stripped of their plumes and left at colonies and, as in Europe, eggs and young left to predators or starvation. The level of disturbance not only caused breeding adults of all species to desert, but they failed to try elsewhere, losing any chance of successful breeding. A repeated theme in Scott's report was the fugitive behaviour of what birds he saw.

As an addendum to Scott's paper, a private letter to him was published in the *Auk* of 1888, confirming what he had written and giving some details of the way the business was operated, particularly that agents of New York firms distributed guns and ammunition to locals for the purpose of shooting birds for plumes. His correspondent recorded that

> *in the spring of 1883, Isodore Cohnfeld, of New York, sent an agent to Punta Rassa, a Mr Kornfeld, with a big lot of guns and ammunition, and he was the first one to inaugurate the crusade against the birds after my arrival here. Kornfeld died at Punta Rassa, but he was replaced by another agent, guns and ammunition were distributed liberally … and the slaughter was carried on.*

In 1884, the plumage collector Batty (who was mentioned by Scott 1887 – see above) started operations in the area, and the following year

> *… had everybody shooting for him. Other parties were in the field also and I received many letters from dealers in New York and Jersey City asking me to buy plumes for them, but I had had enough of it. Bird skins were taken at Myers in the stores in exchange for food and clothing.*

Pearson (1912) carried out a similar exercise to that of Scott, covering 200 miles of the coast of Florida from Tampa to Key West in 1906, and visiting 25 colonies of waterbirds that he had checked 11 years previously. In all this stretch fewer than 12 white egrets were found where they had previously been plentiful or abundant. Pearson also reported that the Great White Egret and the Snowy Egret *Egretta thula* formerly bred

> *… from Oregon and New York in the north, south through Mexico and northern Central America to Patagonia and Chile. Their range, however, in the United States has been greatly restricted. One small colony is reported to be still in existence in eastern Oregon, and it is just possible that there are one or more groups of birds in southern California. The most northern nesting place on the Atlantic coast is in North Carolina, close to the*

southern boundary line. Large areas in Florida, where, in years gone by, the birds were more abundant than in any other area in the United States, are now devoid of either species, except now and then a rare straggler.

Plume hunters were also active in South America, particularly in Venezuela, Colombia, Brazil and Argentina. Del Hoyo *et al.* (1992) recorded that 1.5 million birds were exported from Venezuela in 1898 and that 15,000kg of feathers, mainly from Snowy Egrets and Great White Egrets, were exported from Argentina, Brazil and Venezuela between 1899 and 1912. Doughty (1975) pointed to the difficulty of translating such weights into numbers of birds, but suggested that compromise figures per kilo were 800–1,000 Snowy Egrets or 200–300 Great White Egrets. Accepting these figures, he calculated that 12–15 million Snowy Egrets or 3–4.5 million Great White Egrets would have supplied the weight exported.

The abundant tern colonies along the eastern seaboard of the United States were similarly harassed to the verge of extinction. As already noted, tern colonies were particularly vulnerable to such persecution, being usually sited in areas of easy access such as beaches, sand spits and sand dunes. Bent (1921) noted that five species were particularly sought by the millinery trade for ladies' hats, Gull-billed *Sterna nilotica*, Arctic *Sterna paradisaea*, Common *Sterna hirundo*, Roseate *Sterna dougallii* and Least Terns. A sixth species, the Royal Tern *Sterna maxima*, was also affected for, although not shot for its plumage, the scale of egging to which it was subjected and gross disturbance from the shooting going on all around its colonies caused wholesale colony-desertion.

Shooting for the plumage trade caused a steep decline in the numbers of Gull-billed Terns. The best documented site was Cobb's Island, Virginia, where the terns nested in great abundance in the 19th century, but they had been reduced to 1,000 pairs in 1900 and to eight pairs by 1903. Common and Arctic Terns were nearly exterminated along the entire east coast by the 1890s by a combination of egging and shooting for the plumage trade. The Roseate Tern was treated similarly, and disappeared entirely after 1890 from the coasts of New Jersey and Virginia, where it had once been abundant. All these species started to recover with protection in the early 20th century.

The most persecuted species, however, was the Least Tern, particularly on the coasts of Massachusetts, New Jersey and Virginia. In 1885, Dr Witmer Stone (in Bent 1921) reported seeing two gunners at Cape May with two knee-high piles of dead Common and Least Terns, which they were sending on ice to New York at 12 cents each. Least Terns were astonishingly abundant on the coast of Virginia, but professional collectors for the millinery trade spent most of the nesting season on breeding sites there, killing them in incredible numbers. One gunner reported that 100,000 were sometimes shot in a season. By 1891 all terns had been largely extirpated on the main site of Cobb's Island, and they were virtually extinct by 1903. Least Tern colonies on the coasts of North and South Carolina were similarly hunted, but never completely exterminated (Bent 1921).

Hummingbirds, kingfishers and birds-of-paradise

The preceding notes on herons and terns are an example of the scale and impact of the plumage trade on just two groups of birds. But all brightly coloured birds were also vulnerable; three groups are particularly worth mentioning, the hummingbirds, the kingfishers, and the birds-of-paradise.

Native peoples in Central and South America have long used hummingbird plumage for personal adornment. The Aztecs, for example, kept these birds in aviaries to ensure a supply of feathers for ceremonial purposes. Hummingbirds were much sought by the European and North American fashion trade, and millions of skins were exported from Central and South America in the second half of the 19th century, not only to decorate hats and clothes but also for the manufacture of feather pictures, ornaments and artificial flowers. Some figures of the numbers involved in London have already been given (see p. 197). Together with some further figures given by del Hoyo *et al.* (1999), these total no fewer than 605,000 hummingbird skins passing through the London auctions alone between about 1887 and 1911. Del Hoyo *et al.* also remarked that some 8,000 hummingbird skins were used to make a single shawl in 1905. Some hummingbird species were doubtless hunted to extinction in this slaughter, as a few were first described from skins collected for the fashion trade and never subsequently recorded, although hybridization may have accounted for some of these forms (Hume & Waters 2012). The demand for such plumage was also subject to the vagaries of fashion, and after a run of favour hummingbirds could hardly find buyers at a farthing apiece in London auctions in 1912 (Penry-Jones 1958).

There was a major trade in kingfishers, the scale of which in London is indicated briefly by figures from auction sales there in 1906–1907 (see p. 197). One species much concerned was the White-breasted Kingfisher *Halcyon smyrnensis*, much valued for the blue feathers of its upperparts. Macpherson (1897) reported that there was a considerable export trade in these skins between India and China, and quoted one District Officer in India being attracted to a patch of cobalt blue in the countryside and finding several thousand skins of this kingfisher drying in the sunshine. Macpherson noted that Indian fowlers took kingfishers with nets and decoys, saying that

> … the man has a wild bird in his pocket and walks along the stream until he sees a bird of the same species. He then puts up a small net, which is lightly propped up by a twig and on the other side he pegs down his call-bird. The free bird hardly waits until the man is clear of the net, when it flies straight down to attack the tethered bird, and striking the net it falls down on top of him. The man runs up at once, catches the bird and puts it in his bag. In this way they will clear five or six miles of kingfishers in a single morning.

Kingfishers were also exploited for the fashion trade in Britain, as well as by collectors of specimens. Most of the 19th- and early 20th-century county avifaunas record a significant decline in Common Kingfisher *Alcedo atthis* populations related to collecting for the fashion trade, although it was not yet a rare species, except locally. They were not only shot. Bucknill (1900) noted that they were easy to catch by liming twigs which provided prominent perches along streams. Nelson (1907) quoted Thomas Allis in 1844 that they were also snared and limed in Yorkshire and, on at least one occasion, taken wholesale in nets. Turner (1924) recorded that shore-shooters in Norfolk were paid a shilling each (£1.65 today) for Kingfisher skins, and that they were sometimes caught in flight-nets set for wildfowl. In Europe Kingfishers were hunted by peasants in most countries, mainly taken in nets hung beneath bridges just above the water, into which the birds were driven (MacPherson 1897).

The skins of birds-of-paradise were first recorded being brought to Europe in 1522 by survivors of Magellan's circumnavigation of the globe, as gifts to the King of Spain from the ruler of the Batchian Islands in the Moluccas. However, it is likely that the Portuguese, having reached the Moluccas by 1510, already knew of them and of the existing trade in them in the Far East. Furthermore, it is possible that eastern traders had brought the skins into Europe much earlier; Pierre Belon, writing in 1553, mentions plumes in the adornments of the Ottoman Janissaries (elite soldiery) that could hardly be anything other than those of birds-of-paradise, obtained from Arab traders (Newton 1896).

It is likely that the skins of birds-of-paradise have been traded in New Guinea and from there to what is now southeast Indonesia, the Philippines and mainland southeast Asia for up to 5,000 years; they were certainly valued in Asia more than 2,000 years ago (del Hoyo *et al.* 2009). Del Hoyo *et al.* observed that increased interest in the exploration of the Moluccas, New Guinea and Australia in search of new birds in the first half of the 19th century caused considerable demand for specimens for museums and the millinery trade by the end of the century. During the first decade of the 20th century, hundreds of thousands of birds-of-paradise skins were exported from New Guinea, up to 80,000 each year reaching Europe. Doughty (1975) noted that 155,000 birds-of-paradise skins, especially of Greater Bird-of-Paradise *Paradisaea apoda* and Red Bird-of-Paradise *Paradisaea rubra*, were sold at auction in London alone during 1904–1908. As the millinery trade declined in the early 20th century there was increased demand for live birds-of-paradise for personal, royal or public collections. Commercial trade is now illegal in New Guinea (del Hoyo *et al.* 2009).

Banning the trade

Some good emerged from all this slaughter. Alarm and disgust at the cruelties and impact on populations of the plumage trade led to the foundation of several modern conservation movements, with the formation of the Society for the Protection of Birds (originally known as 'The Plumage League', and, from 1904, the Royal Society for the Protection of Birds) in the United Kingdom and the Audubon Societies in the United States in the late 19th and early 20th centuries.

Doughty (1975) gave an excellent and detailed account of the struggles to ban the trade. Reading his account, one is struck by the familiarity of the arguments trotted out in defence of the indefensible. As in similar circumstances today, emphasis was continually placed on the value of the trade and the employment it provided. The value of the trade was certainly considerable. In 1912, for example, egret plumes were selling at auction in London for £12–17s–6d per ounce (about £1,171 today); they were then more valuable than their weight in gold. Evasions and outright lies were told about the management of the trade, particularly on the way plumes were harvested. The ludicrous claims that these were all gathered as moulted feathers after breeding was skewered by Pearson (1912), but it was nevertheless still believed in some circles. Pearson destroyed this spurious argument by quoting numerous affidavits on how the hunting was done from plume-hunters and travellers familiar with the trade.

Public outrage eventually forced the imposition of a ban, and the trade was made illegal in 1921. One strong impression I am left with by Doughty's (1975) account of the battles protectionists had to fight to be believed by and get action from governments is the error of relying too heavily upon politicians for the protection of wildlife. Subject to pressure from many vested interests, politicians too often seek unsatisfactory compromises, and can often have different agendas to the rest of us. It should also be noted that the impact of World War I on fashion, the fashion trade and the status of women may have had as much to do with the demise of the plumage trade as the legislation did.

CHAPTER 12
Some conclusions

The over-exploitation of birds for food, feathers and for collections put a severe strain on the populations of a number of species in the 19th and early 20th centuries. Examples can readily be found in the preceding chapters among the herons, wildfowl, waders and seabirds. But the situation was rarely as clear-cut as one would wish. Three major forces contributed to changes in bird populations and distribution over the period covered by this book – climatic change, habitat changes, particularly those arising from the spread of and changes in agriculture, and persecution and exploitation.

There have been two major periods of climatic change in this period. The first was a long period of climatic deterioration known as the Little Ice Age, which lasted from about the mid-14th century to the mid-19th, and was at its most intense from the late 16th century to the early 18th (Fagan 2000). The second was a period of climatic amelioration which started around 1850 and continues to this day. Neither trend was continuous, and there have been marked fluctuations in both.

How far the climatic changes of the Little Ice Age affected bird populations is largely unknown in the absence of detailed records. Assumptions can be made on the basis of modern experience. But these may not be valid where they occurred against a background of deterioration and loss of habitat, particularly to agriculture, and over-exploitation. The Little Egret *Egretta garzetta* provides a good example of this. The increasing frequency of cold winters in the Little Ice Age would almost certainly have affected populations (Voisin 1991) but so too, probably, did over-exploitation for food, which was a matter of status rather than necessity (see p. 52). But Arctic and sub-Arctic species would certainly have been forced steadily southwards in the face of the advancing ice. Such a shift in distribution may explain why Dotterel *Charadrius morinellus* did not appear in records such as household accounts and market lists in Britain before the 16th century. Species we now regard as southern European may have shifted away from northwest Europe in the colder conditions; one such bird is the Hoopoe *Upupa epops* (Fisher 1966).

The deterioration in climate certainly also contributed to important changes in farmland habitats, perhaps helping to drive the increasing enclosure of the commons in England from the mid-18th century. Difficult weather conditions for arable agriculture were frequent in the second half of that century, and contemporary authors had much to say about the poor state of arable farmland, usually ascribing it to deficiencies in the open-field agricultural system. But it is quite probable that the physical difficulties of adequately cultivating the clay soils of the open field parishes in increasingly unfavourable weather conditions may actually have been beyond the technical capacity of 18th-century farmers (Shrubb 2003). The enclosure of the wastes and commons brought large areas of light, warm, well-drained and easily cultivated soil into the system, the fertility of which was raised and maintained by the rotation farming methods also being applied (Shrubb 2003). But it also involved major habitat changes for birds and, for example, the eventual loss of lowland populations of Black Grouse *Tetrao tetrix* in England largely stemmed from it (p. 98). The clayland areas of the open fields increasingly went to pasture, a change that would also have affected bird populations. This was also a largely British problem. Agricultural change in the rest of Europe was a much slower process, and in many areas became a feature of the 20th century, with the development of the European Union and its Common Agricultural Policy (Shrubb 2007).

There was also extensive drainage of major wetlands in Britain from the mid-18th century, a change that again was much slower elsewhere in Europe except, possibly, in Denmark. This change resulted in the elimination or reduction of fen species such as Bittern *Botaurus stellaris* and Black-tailed Godwit *Limosa limosa*. Burton (1995) pointed out that much apparently suitable habitat remained in East Anglia for these species well into the 19th century, and suggested that climatic deterioration was an important cause of their loss. This may be so, but arable agriculture in the Fens sought to lower water tables by up to a metre, which may have particularly affected species such as Black-tailed Godwit or those needing wet reedbeds such as Bitterns (Shrubb 2003). Over-exploitation certainly also contributed.

In a wider perspective, the clearance of woodland for agriculture undoubtedly favoured open-country species. The Lapwing *Vanellus vanellus* (Shrubb 2007) and the Grey Partridge *Perdix perdix* provide examples of this. The decline of human population and of arable agriculture in favour of pasture throughout Europe after the Black Death apparently encouraged an expansion of the Great Bustard *Otis tarda* in northwest Europe and into Britain (Shrubb 2011). What effect the loss of woodland had is largely unknown, but the Capercaillie *Tetrao urogallus* disappeared from mediaeval England and 18th-century Ireland and Scotland largely as a result (see p. 100).

A particular feature of the exploitation of birds has been their killing (including egging) as spring migrants and in the breeding season. This has been a recurrent theme throughout this book, and continues in some parts of the world, particularly Greenland, where the Inuit are presently engaged in destroying their wildlife in the name of maintaining their traditional culture – with high-powered motor boats and modern firearms. Many examples of the impact of such over-exploitation are given in the preceding chapters, which are summarised in Table 12.1.

Great Crested Grebe	Shooting for the plumage trade seriously reduced numbers in Britain from mid-19th century. Recovered with protection from late 19th century (Ch. 11).
Gannet	Colonies in the Nearctic seriously reduced by over-exploitation. In Britain one colony, at Lundy, destroyed by persecution, the birds moving to Grassholm off Pembrokeshire; the population otherwise appeared stable. But a major increase in numbers with protection in the 20th century suggests that exploitation held the population at an artificially low level (Ch. 8).
Bittern	Shooting, combined with drainage of major wetlands, extinguished breeding populations in a large area of northwest Europe. It may be that shooting winter visitors in Britain prevented recolonisation in the 19th century. Protection and habitat management have led to some recovery in the 20th century (Ch. 4).
Night Heron	Serious population decline in Europe into the 19th century caused by land drainage exacerbated by overexploitation for food. Some recovery in the 20th century (Ch. 4).
Egrets	The plumage trade world-wide caused massive destruction of breeding colonies (Ch. 11). Little Egret disappeared from northwest Europe in the early modern period probably as a result of over-exploitation for food combined with climatic deterioration. Substantial range expansion in the second half of the 20th century probably results as much from total protection as climatic change (Ch. 4).
Spoonbill	Combination of land drainage and the taking of eggs and young for food led to its virtual extinction in western Europe. Now recolonising Britain (Ch. 4).
Whooper Swan	Over-exploitation for skins combined with habitat loss caused serious declines in Europe (Ch. 5). Has recovered with protection in Scandinavia (Hagemeijer & Blair 1997) but continues to decline in Russia despite protection (Flint *et al.* 1984).
Wild geese	Netting and shooting of moulting birds for food and feathers and egging in spring caused serious population declines in Soviet Russia in Greylag, Greater White-fronted, Bean and Brent Geese (Ch. 5). Greenland White-fronted Geese badly affected recently by the shooting of spring migrants in Iceland, now banned (Holt *et al.* 2009). Brent Geese in Svalbard seriously reduced by egging in the late 19th and early 20th centuries. Protection in the 20th century has seen major population recoveries (Ch. 5).
Teal, Pintail, Garganey, Shoveler	Spring shooting and trapping saw significant declines in Britain, which were reversed by protection from the 1880s (Ch. 5).
Sea ducks	In the Baltic region, shooting and netting birds on spring migration and unrestricted exploitation for eggs caused serious population declines by the late 19th century (Ch. 5).
Common Eider	Unrestrained spring and summer shooting and egging in much of the north Atlantic area, particularly in the northeast United States, Canada and Greenland, brought many populations to the verge of extinction by the end of the 19th century (Ch. 5). This remains a problem in Greenland today, but elsewhere populations have recovered with the establishment of reserves, banning egging, limiting spring shooting and, in the Baltic, probably with improved feeding conditions (Hagemeijer & Blair 1997).

Black Grouse	Unrestricted shooting, particularly at leks, in Scandinavia caused serious population declines in the 19th century (Ch. 6). Commons enclosure in the 19th century largely instrumental in the loss of lowland populations in England, and the loss of lowland heathlands in Denmark, Sweden and the Netherlands has led to similar declines there since the 18th century.
Quail	Excessive netting and shooting of spring migrants around the Mediterranean, particularly from the late 19th century, caused major declines in European breeding populations, from which the species has not recovered (Ch. 6).
Crane	Persecution and overexploitation in the breeding season and on spring migration from the mediaeval period caused the loss or reduction of western European populations. With protection, recovered in the second half of the 20th century and now recolonising England (Ch. 4).
Great Snipe	Netting at leks and excessive shooting of autumn migrants in Sweden in the mid-19th century led to steep population declines, from which the population has not recovered (Ch. 7).
Stone Curlew	Virtually extinguished by egg-collecting on Dungeness in Kent (Ch. 10). Agricultural change in lowland England led to the loss of many farmland populations in the 20th century, but there has been some recent recovery with conservation measures.
Kentish Plover	Egg-collecting and shooting breeding birds for collections virtually exterminated the British population, which was confined to the coasts of Sussex and Kent, by the end of the 19th century (Ch. 10).
Dotterel	The shooting of large numbers of spring migrants in Britain and northwest Europe caused major declines in breeding populations. Migrants disappeared completely from many traditional areas in Britain from the late 19th to the mid-20th centuries. With protection, the British population recovered markedly in the second half of the 20th century (Ch. 7).
Lapwing	A major northward expansion of range in Europe between the late 19th century and the 1970s resulted from a combination of climatic amelioration, improving winter survival, and declining persecution, particularly of egging, improving productivity (Ch. 7; Shrubb 2007). A sharp decline in Britain due to agricultural change since.
Ruff	A combination of land drainage, trapping birds at lek and collecting eggs and specimens caused the loss of the British population (Ch. 7). Recent attempts to recolonise Britain have not been successful.
Black-tailed Godwit	Certainly over-exploited in the Netherlands in the mid-19th century, leading to serious population decline (Stevenson 1870). Recovery since, but agricultural intensification now a problem (Hagemeijer & Blair 1997).
Black-headed Gull	Increasingly excessive egging in England caused a marked decline in the population in the 19th century. The population started to recover with the passing of the Birds Protection Acts in the 1880s (Ch. 8).
Black-legged Kittiwake and terns	Mindless summer shooting for 'sport' and the plumage trade damaged British populations (Chs. 8 & 16). Tern populations were particularly seriously damaged by shooting for the plumage trade along the whole eastern seaboard of the United States (Ch. 11). Populations have recovered with protection in the 20th century, although problems with pollution and over-exploitation of fisheries now affect this (Hagemeijer & Blair 1997).

Auks	Gross over-exploitation for meat, eggs and sometimes feathers seriously reduced major populations in the northeast United States, Labrador and Greenland; many colonies were virtually wiped out. Over-shooting remains a serious conservation problem in the latter two. Brünnich's Guillemot similarly reduced in the Russian Arctic. Summer shooting for 'sport' reduced numbers in many English auk colonies (Chs. 8 & 11). Populations have recovered with protection.
Great Auk	Exterminated for meat, feathers and oil (Ch. 8).
Goldfinch	A serious decline in Britain in the 19th and early 20th centuries resulted from large-scale trapping for the cagebird trade combined with major habitat changes in farmland. The population has recovered with protection (Ch. 10).
Bullfinch	Decline in England in mid-19th century resulting from trapping for cagebirds and persecution. Numbers recovered from the early 20th century with protection (Ch. 10).
Passerines	Egg-collecting, combined sometimes with specimen collecting, significantly damaged populations of some scarcer species, eg. Bearded Tit, Dartford Warbler and Marsh Warbler, in the 19th and early 20th centuries. The first two have recovered with protection.

Table 12.1. *Summary of the impact of over-exploitation and other factors on bird species in Britain and Europe (for seabirds the north Atlantic) prior to the early 20th century.*

The populations of many of these species recovered in the 20th century with protection, the implementation of proper close seasons, and sometimes with the creation of nature reserves. Some populations, particularly of the herons, Spoonbill *Platalea leucorodia* and Common Crane *Grus grus*, have recolonised breeding ranges not occupied for centuries. The present expansion of the Little Egret *Egretta garzetta* is usually ascribed to climatic change, but the scale of its destruction for the plumage trade from the mid-19th century makes it likely that the total protection it now enjoys is the predominant cause.

Whilst in the absence of detailed counts it is difficult to know how far exploitation from the mediaeval period onwards affected seabird colonies in Britain and northwest Europe, protection since the late 19th century certainly allowed or contributed to an expansion in the 20th century in the range and numbers of Gannets *Morus bassana*, Common Guillemots *Uria aalge*, Razorbills *Alca torda*, Black-headed Gulls *Larus ridibundus*, Kittiwakes *Rissa tridactyla* and possibly Cormorants *Phalacrocorax carbo* and Shags *Phalacrocorax aristotelis*. Puffins *Fratercula arctica* tended to decline in the early 20th century despite a drop in exploitation, probably from specific causes such as the invasion of breeding colonies by rats. Large gulls, particularly Herring Gulls *Larus argentatus* which were much exploited for eggs (Cott 1953), have declined recently after a significant increase earlier in the 20th century (Lloyd *et al.* 1991).

There seems to be no certainty that sea duck populations in Scandinavia and the Baltic benefited from protection in the early 20th century, except for Goldeneye *Bucephala clangula*, which have increased with the increased provision of nest boxes, and Goosanders *Mergus merganser*. Some, such as Velvet Scoter *Melanitta fusca*, are still hunted at sea and in the breeding season and others, Long-tailed Duck *Clangula hyemalis*, for example, are now affected by such factors as oil pollution (Hagemeijer & Blair 1997).

Nor is the table exhaustive. I have not discussed groups such as pigeons and doves, nor raptors and owls which were heavily persecuted by game-preserving interests, and have been covered by Lovegrove (2007). But the taking of raptors for falconry almost certainly restricted populations in the mediaeval and early modern periods. Demand from Middle Eastern countries continues to exert pressure on the populations of some, particularly Gyr *Falco rusticolus*, Saker *Falco cherrug* and Peregrine Falcons *Falco peregrinus* (see p. 31).

Most waders were heavily exploited for food, but evidence that this affected populations other than of the species listed in Table 12.1 is lacking, except perhaps for Pied Avocet *Recurvirostra avosetta*, which was also much sought by collectors. It is also difficult to believe that the enormous numbers of passerines taken for food and the cagebird trade had no impact on populations. But, except for Goldfinch *Carduelis carduelis*, Bullfinch *Pyrrhula pyrrhula* and probably Linnet *Carduelis cannabina*, there is little evidence for this in the 19th-century literature, at least for Britain, although this may reflect a lack of numerical data rather than the true situation.

Table 12.1 is largely confined to Britain and Europe, and it is worth noting that although many species suffered significant or serious declines, at least regionally, none bar the Great Auk *Pinguinus impennis* was extinguished, although the egrets may have come close to this in the plumage-trade era.

Europe compared with North America

The situation in North America differed considerably. In addition to the Great Auk, four species were exterminated by gross over-exploitation – the Eskimo Curlew *Numenius borealis*, Heath Hen *Tympanuchus cupido*, Passenger Pigeon *Ectopistes migratorius* and Carolina Parakeet *Conuropsis carolinensis*. The Labrador Duck *Camptorhynchus labradorius*, also extinct, may have been affected by shooting, as it appeared from time to time in the markets, but there is no real understanding of why it disappeared. Greenway (1967) noted that these birds were all specialised breeding endemics and therefore vulnerable to exploitation. The Trumpeter Swan *Cygnus buccinator* and Whooping Crane *Grus americana* were also brought to the brink of extinction by exploitation.

Greenway (1967) also observed that the history of the recent extinction of birds in North America coincided so closely with the penetration of the continent by European settlers that it could not be doubted that they represented cause and effect. One particular feature of the over-exploitation of gamebirds, waders and wildfowl was large-scale, indiscriminate shooting for the market, irrespective of season; significant numbers of these birds were exported to Britain. No check was placed on this by close seasons, game preserves or similar devices. These were not introduced until the late 19th century, when proper close seasons and bag limits were imposed. In this, North America differed sharply from the situation in Britain and Europe for, as noted at the start of this book, in Britain and Europe the land was fully owned and controlled by the mediaeval period. Social regulation also exercised control over exploitation in the interest of conserving stocks, or by limiting those entitled to hunt.

Prehistoric peoples, of course, were not so constrained. Serjeantson (2001) discussed the archaeological evidence for the exploitation of the Great Auk and the Gannet in Britain in the prehistoric era. He concluded that the distribution of both species had been limited to more inaccessible islands and offshore stacks by exploitation and disturbance of mainland and inhabited island sites by the first millennium AD. It is probable that other colonial seabirds were similarly affected. The restrictions controlling exploitation in Britain and Europe showed a marked tendency to break down in the 19th century; in Britain, for example, with the loss of close seasons for wildfowl under the 1831 Game Reform Act and the persecution of seabird colonies for 'sport', and in Europe with the loss of aristocratic privilege in the political upheavals of the mid-19th century. There seems little doubt that such factors made an important contribution to the increased impact of exploitation and persecution of birds obvious in the 19th and early 20th centuries. But it would be wrong to suppose that the over-exploitation of bird populations in the region is a thing of the past; Greenland provides an ongoing modern example.

The present situation in Greenland

There are many examples from history cited in this book that show that hunters had no instinctive feeling for nature and conservation, but would hunt birds to extinction unless prevented, often because they regarded such populations as an infinite resource (Serjeantson 2001, Nicholls 2009) or had reduced hunting to a purely commercial activity. Protests to hunters about the damage caused, for example by the plumage trade, produced the counter argument that 'If I don't kill them, someone else will, so I might as well have my share'. Such strictures apply also to indigenous native peoples, of which the Inuit in Greenland are a prime example. Much has been written about the sustainable hunting of the Inuit, which, as Hansen (2002) has pointed out, is politically correct bunkum. Early European explorers in Greenland wrote of mounds of rotting meat as common features of Inuit settlements (Hansen 2002), hardly an indication of a hunting culture in tune with nature. Such waste continues. Hansen noted that the Common Eider was one of the best documented examples of how wanton over-exploitation wrecked populations in the past. The decimation of the population of this duck was largely complete 160 years ago, and large colonies have never re-established themselves over most of coastal Greenland in the face of continued over-exploitation.

Sale & Potapov (2010) noted that over 100,000 significant bird colonies were estimated to exist in Greenland in 1845, of which only a few thousand remain today as a result of unrestrained shooting and egging. Hansen (2002) indicated that this was a continuous process over the whole period, undermining arguments about sustainable hunting in the past. Hansen also quoted F. Sejersen, an anthropologist at Copenhagen University, on the myth of the sustainable exploitation of food resources by the Inuit in the past. This argument rested on three points – the Inuits' view of nature, their limited populations, and limited technology. For the first, Sejersen quoted examples from the Arctic arguing that traditional hunters and trappers never considered species' survival, but only their own. For the second and third he pointed out that although these factors might limit the numbers of animals taken there was no guarantee of this, as the destruction of large mammal populations in North America in the prehistoric era (Yalden 1999) surely makes clear.

Protective legislation for birds was passed in Greenland in 2001 but, under pressure from Greenlanders, was cancelled in 2004. The pressure came from professional hunters and from ordinary Greenlanders who wished to hunt for pleasure or, as they put it, as 'a way of keeping in touch with their roots' (Sale & Potapov 2010). I have watched a similar slaughter in Barrow, north Alaska, when the wildfowl and waders were arriving in spring; nobody bothered to pick up the birds shot. They didn't really have to, as they bought their food in the supermarket. An unwelcome side effect of a century of shooting in this area is the fact that lead shot has accumulated to such an extent that ingestion by female Spectacled Eiders *Somateria fischeri* has reduced their annual survival by 35%, from lead poisoning alone (Studebaker 2012).

Hansen (2002) noted that the present-day hunting culture in Greenland had become a 'subsidy culture' grosser than any EU farming subsidy. He wrote that 'professional hunters, dinghy fishermen and sheep farmers annually receive millions in subsidies … without any prospect of Greenland society making reasonable use of what these groups catch or produce'. Hunters in Greenland annually kill birds and animals sufficient to supply the whole of Greenland society with food from traditional sources. But more and more Greenlanders prefer to eat imported foods, and three quarters of the population's meat needs are met by beef and pork imported from Denmark. Thus the wastage from kills in Greenland is enormous, with thousands of tonnes of meat from killed animals and birds never used. Egging at seabird colonies in Greenland also remains a serious problem, as noted for Arctic Tern *Sterna paradisaea* and Brünnich's Guillemot *Uria lomvia* in Chapter 8. The Danish and Greenland Governments have supinely allowed Greenlanders to destroy their wildlife. They may find one unwelcome result of this to be a serious impairment of their tourist trade, attracted, as it is, by what remains of the wildlife.

Newfoundland and Labrador

Similar problems have arisen elsewhere in the northwest Atlantic. Del Hoyo *et al.* (1996) noted that one of the most serious recent assaults on alcids has been the winter shooting of guillemots off the coast of Newfoundland and Labrador, as a result of a change from traditional subsistence hunting in the 1940s to a massive black market and recreational sport hunt, without any limits. Del Hoyo *et al.* noted that the killing of guillemots in the northwest Atlantic has become the most pressing seabird conservation issue in the northern hemisphere today – an unwelcome case of history repeating itself.

We can do nothing about the mistakes of the past, except learn from them. But it is not too late to bring some governments to their senses, and stop further needless slaughter. Otherwise, in the case of northern seabirds, some of which now face serious threats to their survival, political correctness and weak government will extend into a failure to adequately control oil and gas exploration, as climatic changes opens up new areas for exploration.

Appendix 1

Present status of Grey Heronries in England and Wales described as 'immemorial' or 'ancient' by Nicholson 1929, which were still active in 1954. Foundation dates are from Nicholson 1929.

Berkshire	The colony at Coley Park was founded in 1829 but was deserted after the 1962–63 winter (Standley *et al.* 1996).
Breconshire	The Sennybridge colony, which existed before 1884, was last occupied in 1995, with four nests (Breconshire Bird Report).
Ceredigion	The ancient site at Llanllyr near Talsarn had 9 pairs in 2007. That at Highmead, Llanbyther (probably founded before 1750) had 18 in the same year. The colony at Llidiardau, Llanilar, which may have been an ancient site and shifted to Maesllyn in 1990, had 21 nests in 2007 (Roderick & Davis 2010).
Cornwall	The colony at Trenant Park, which was founded before 1850, was still occupied in 2006, with six nests (Cornwall Bird Report).
Cumberland	The ancient site at Muncaster, which was founded before 1621, had 14 nests in 2010. The immemorial site at Edenhall had no nests in 2003; that at Greystoke was then extinct (John Marchant pers. comm.).
Devon	The ancient colony at Sharpham had declined to a single nest by 1980 and was extinct by 2007. Of the immemorial sites Killerton was extinct by 1980, Warleigh and Puslinch by 2007 and Hallwell House and Orcheton Wood were apparently not recorded after 1954. The colony at Arlington Court held 28 nests in 2007 (Rogers 2008, Tyler 2010).
Dorset	The colony at Crichel Park, which was founded before 1835, was destroyed in 1962 (Green 2004).
Essex	Brightlingsea/St Osyth – although listed as immemorial by Nicholson (1929), both Christy (1890) and Glegg (1929) record it as active in the first half of the 18th century. In 1872 the colony moved from Brightlingsea to St Osyth, where still extant. The colony at Wanstead Park was certainly active well before 1834 and shifted to Walthamstow Reservoir around 1914, where still extant, with a maximum of 110 nests during 2000–2004 (Wood 2007).
Glamorgan	The Hensol colony, which was founded before 1872, was still occupied in 2010, with 15 nests. The Penrice colony was abandoned after 1993 (Heathcote *et al.* 1967, Glamorgan Bird Report)
Hampshire	The colony at Sowley Pond, which Nicholson (1929) recorded as active well before 1901, was still occupied in 2007, with 11 nests (Hampshire Bird Report).
Herefordshire	The colony at Berrington Hall, first recorded in the 1880s, had 26 nests in 2006 (Herefordshire Bird Report).

Kent	The ancient colony at Chilham, which was in existence before 1280, had three pairs in 1994 and was presumably extinct thereafter, as there were no subsequent reports (Kent Bird Report).
Lancashire	The colony at Ashton Hall, which was planted between 1800 and 1810 by the Duke of Hamilton (Oakes 1953) had five pairs in 2003. The colony at Ince Blundell, which was founded before 1863, was still active in 2003 and that at Claughton (founded before 1874) had 50 pairs in that year. The ancient site at Rusland Moss and that at Scarisbrook Hall both appeared to be extinct by the late 20th century (Lancashire Bird Report).
Leicestershire	The colony at Stapleford Park was first noted in 1840 with six nests. There were 16 in 1988 and between three and eight in the early 21st century (Fray *et al.* 2009, Leicestershire Bird Report).
Lincolnshire	The colony at Muckton Wood, which was founded before 1870, was still occupied in 1992, with 12 nests, after which it was deserted due to disturbance (Lincolnshire Bird Report).
Norfolk	The immemorial Islington colony had 52 nests in 2003. The ancient site at Reedham apparently abandoned after 1977 although satellite colonies recorded. Reedham itself reoccupied in 2008 and four nests in 2009 (Taylor *et al.* 1999, Norfolk Bird Report).
Northamptonshire	The colony at Althorp was unoccupied for the first time since probably 1567 in 1994; there were 43 nests in 1991. The Milton Park colony, which was probably founded in 1819, appeared to be extinct by the 1980s (Northants Bird Report).
Northumberland	The colony at Chillingham has not been specifically recorded since the 1970s (Northumberland Bird Report), and must be presumed extinct.
Nottinghamshire	The colony at Stanford, which was probably founded before 1830, had seven nests in 2007. Those at Colwick and East Stoke were both extinct in the 1950s (Dobbs 1975, Notts Bird Report).
Oxfordshire	The immemorial colony at Buscot, formerly in the pre–1974 county of Berkshire, had 26 pairs in 2002 and was still occupied in 2007 (Oxon Bird Report).
Shropshire	The colony at Halston Hall, which dates back to at least the 1820s, had 23 nests in 2000 and was still occupied in 2005. That at Attingham/Sudborne apparently became extinct in the late 1950s (Shropshire Bird Report).
Somerset	The colony in Somerton Wood, which was founded before 1878 and was probably an ancient site, had 22 nests in 2004. The ancient site at Brockley Park shifted to Cleeve Hill after 1965 but was not recorded in 1988 or 2004. The ancient site at Allers Wood/Pixton, which was probably founded before 1545, was extinct in 1959 (Ballance 2006).
Staffordshire	The ancient colony at Bagots Wood had more than 60 nests in 2003. That at Aqualate Mere, which was in existence in 1686 (Smith 1938) had more than 40 in 2003. The colony at Consall Wood was extinct by the early 1980s (Harrison & Harrison 2005).
Suffolk	The Blackheath colony, which was founded before 1871, had 10 nests in 2003. The ancient site on the Orwell estuary, which existed before 1600, has moved several times around the estuary and is now at Wolverstone Wood, where there were 16 nests in 2003 (Wright 1986, Suffolk Bird Report).

Surrey	The history of the Windsor Great Park colony goes back to at least 1607 and there were seven nests in 2003 (Wheatley 2007).
Sussex	There has been a substantial heronry in the triangle formed by Rye, Brede and Beckley since at least 1297. Originally at Udimore, it has moved at least four times since 1840. The present site at Leasam near Rye was first colonised about 1939 and held 30 nests in 2004. There has been a colony in the Herstmonceux/ Wartling area since at least the mid-16th century. It has moved at least four times, and died out briefly in 1948. The present site at Wartling held 13 nests in 2004. The colony at Parham was founded at Michelgrove in 1810 and moved to its present site during 1826–32. It held 23 nests in 2004 (Shrubb 1979, Sussex Bird Report).
Warwickshire	The colony at Combe Abbey continued to grow in 2003. That at Warwick Park was extinct between 1974 and 2001 but had re-established itself in 2003. Ragley Park had 8–10 nests in 2007 (Harrison *et al.* 1982, Harrison & Harrison 2005).
Westmorland	The ancient colony at Dallam Tower, which Nicholson (1929) noted shifted in 1775, was still occupied in the early 21st century, with 27 nests in 2011 (John Marchant, pers. comm.).
Wiltshire	The colony at Bowood Park, which was founded well before 1852, was still active in 2000. Although the ancient site at Longleat was not recorded in 1954, it was still occupied between 1990 and 2000 (Wiltshire Bird Report).
Yorkshire	The ancient site at Flasby had shifted to Coniston Cold in 1971, when it had 71 nests. Although still flourishing in 1986, it was not reported thereafter. The colony at Harewood Park was extinct after 1962 and that at Gillings Wood was apparently not reported after the 1950s (Mather 1986, Yorkshire Bird Report).

Appendix 2

Duck decoys in England and Wales; none was ever completed in Scotland. The information given comprises county, total sites, site name, 10-km square, date constructed if known, and date abandoned if known, in that order. References are given below each county list.

Bedfordshire (three sites)

Site name	10-km square	Date built if known	Date abandoned if known
Tempsford	TL15	early 18th century	? mid 19th century
Houghton	TL04	17th century	? late 18th century
Little Barford	TL15	?	?

The decoy at Houghton was a cage decoy. No record exists of that at Little Barford except the name on the tithe map of 1844. References: Steele-Elliott 1936, Key 1955, Fadden 2011.

Berkshire (one site)

Site name	10-km square	Date built if known	Date abandoned if known
Aldermaston Park	SU56/66	?	Probably early 19th century

Reference: Payne-Gallwey 1886.

Buckinghamshire (four, or six sites if Wotton counted separately)

Site name	10-km square	Date built if known	Date abandoned if known
Boarstall	SP61	1691–1697	Restored 1960s and used for ringing
Winchendon	SP71	?	Probably late 18th century
Wotton	SP61	before 1820	1875
Claydon	SP72	?	Late 18th century

There were three small decoys at Wotton, built as house decoys. References: Payne-Gallwey 1886, Kear 1990.

Cambridgeshire (five sites)

Site name	10-km square	Date built if known	Date abandoned if known
Holme Fen	TL28	1815	1834 or 1844
Chatteris	TL38/48	early 17th century	destroyed *c.*1660
Leverington	TF41	before 1760	*c.*1854
Whittlesey	TL29	?	*c.*1852

Wentworth-Day also noted that a decoy was known to have existed on Burnt Fen (TL58) north of Ely, but nothing else was known of it. References: Payne-Gallwey 1886, Wentworth-Day 1954.

Cheshire (one site)

Site name	10-km square	Date built if known	Date abandoned if known
Dodleston	SJ36	before 1634	?

References: Coward & Oldham 1900, Southwell 1904.

Cornwall (one site)

Site name	10-km square	Date built if known	Date abandoned if known
Trengwainton	SW43	1819–20	by 1835

A cage decoy. Reference: Payne-Gallwey 1886.

Derbyshire (one site)

Site name	10-km square	Date built if known	Date abandoned if known
Hardwick Hall	SK46	?	Recently restored by the National Trust

A cage decoy. Reference: Payne-Gallwey 1886.

Dorset (three sites)

Site name	10-km square	Date built if known	Date abandoned if known
Abbotsbury	SY58	1655	still worked for ringing
Morden	SY99	?	1856
?	SY78/88	1720s	?

The last of these three decoys was described by Defoe. It has generally been considered to be the Morden decoy, but Defoe's account is quite clear. He saw this newly constructed decoy *en route* from Wareham to Weymouth, riding 'in view of the sea' on downland. He went from Weymouth north to Dorchester. Morden is six miles north of Wareham on heathland and does not fit Defoe's description. Payne-Gallwey also drew attention to a decoy started at Wimbourne (SU00) but never completed. References: Payne-Gallwey 1886, Mansel- Pleydell 1888, Furbank *et al.* 1991 (for Defoe).

Essex (37–43 sites)

Site name	10-km square	Date built if known	Date abandoned if known
South Hall, Paglesham	TQ99	?	*c.*1790
Southminster	TQ99	?	?
Grange, Tillingham Marsh	TM00	?	still working 1936
Marsh House, Tillingham	TM00	end 18th C	worked occasionally 1933
Glebeland	TM00	?	*c.*1820
East Hall, Bradwell	TM00	?	*c.*1840s
West Wick, Bradwell	TM00	?	*c.*1840s
Ramsey Island	TL90	existed 1772	?
Steeple, Canney Marsh	TL90	1713 enlarged 1721	?
Mayland	TL90	?	?
Latchington	TL80	before 1760	? late 18th century
Northney Island, Maldon	TL80	after 1663	? late 18th century
Goldhanger No. 1	TL80	before 1735	1870
Goldhanger No. 2	TL90	?	*c.*1840s
Goldhanger No. 3	TL80	?	?
Joyce's Decoy	TL90	?	1865
Gore Decoy, Goldhanger	TL90	?	*c.*1830s
Skinner's Wick	TL90	?	1860
Bohun's Hall	TL90	?	early 19th century
Old Hall No. 1	TL91	?	1890
Old Hall No.2	TL91	?	early 19th century
West Mersea	TM01	before 1807	1870
Villa Farm	TM02	?	?
Lion Point	TM11	1860	never worked
Kirby-le-Soken	TM22	?	?
Old Moze Hall	TM22	?	1841
Great Oakley Hall	TM22	?	1840s
Horsey Island	TM22	?	1840
Dovercourt	TM23	?	1820s
Roydon Hall	TM13	?	*c.*1840s

Site name	10-km square	Date built if known	Date abandoned if known
Jacques Hall Old Decoy	TM13	?	?
Jacques Hall	TM12	?	*c.*1840s
Pond Hall	TM13	before 1754	?
Wormingford	TL93	?	?
Felsted	TL71	in existence 1679	?
Rolls Farm	TL90	?	?
Rolls Farm No.2	TL90	?	?

Christy (1903) stated that there was probably a decoy near Fobbing (TQ78) and that one was marked higher up the Thames on an old map of Essex once owned by F. J. Stubbs. Christy (1890) also suspected the former existence of others on Osey Island (TL90) and at Mundon (TL80) from information supplied by Mr E. A. Fitch, but he could find no trace of them. He also suggested that there may have been two more in Great Stambridge parish (TQ89), marked on old maps as Old Pool and New Pool. References: Christy 1890, 1903, Glegg 1943, 1944.

Glamorgan (one site)

Site name	10-km square	Date built if known	Date abandoned if known
Llanrhidian	SS49	?	?

Reference: Payne-Gallwey 1886.

Gloucestershire (two sites)

Site name	10-km square	Date built if known	Date abandoned if known
Berkeley Castle (2)	SO70		

The new decoy was constructed in 1843 and is still used by the Wildfowl and Wetlands Trust for ringing and demonstrations. It was constructed to replace an earlier decoy abandoned when the Gloucester and Sharpness canal was built. Reference: Kear 1990.

Hampshire (three sites)

Site name	10-km square	Date built if known	Date abandoned if known
North Stoneham	SU41	*c.*1806	*c.*1874
Bournemouth	SZ19	?	?
New Forest, Ipley Farm	SU30	?	mid-18th century (or earlier)

Reference: Payne-Gallwey 1886.

Herefordshire (one site)

Site name	10-km square	Date built if known	Date abandoned if known
Shobdon Park	SO36	?	early 19th century

Reference: Bull 1888.

Hertfordshire (one site)

Site name	10-km square	Date built if known	Date abandoned if known
The Hoo, Welwyn	TL21	1870	in use 1880s

A cage decoy. Reference: Payne-Gallwey 1886.

Kent (three sites)

Site name	10-km square	Date built if known	Date abandoned if known
Grovehurst	TQ96	1736	1866
Kemsley	TQ96	?	1780s
High Halstow	TQ77	?	?

References: Payne-Gallwey 1886, Ticehurst 1909.

Lakeland (Cumberland and Westmorland) (two sites)

Site name	10-km square	Date built if known	Date abandoned if known
Muncaster Castle	SD19	?	end 18th century
Lowther Castle	NY52	?	end 18th century

Reference: MacPherson 1892.

Lancashire (three sites)

Site name	10-km square	Date built if known	Date abandoned if known
Hale	SJ48	at least 1735 and probably 1631	still in use for ringing
Orford Hall	SJ68	?	c.1754
Martin Mere	SD41	before 1818	?

References: Payne-Gallwey 1886, Mitchell 1885, Kear 1990.

Lincolnshire (forty sites)

Site name	10-km square	Date built if known	Date abandoned if known
Keadby	SE81	?	after 1839
Ashby	SE80	early 19th C	after 1886
Twigmoor	SE90	?	?
Broughton	SE91	?	?
North Kelsey	TF09	?	?
Freshney Bog	TA20	?	?
Farlesthorpe Fen	TF47	?	?
South Carleton	SK97	?	?
Burton	SK97	?	?
Skellingthorpe	SK97	late 17th century	1840
Nocton Fen	TF06	?	1820
Timberland	TF15	?	?
North Kyme	TF15	?	?
South Kyme	TF14	18th century	?
Sempringham Fen	TF13	?	?
Aslackby	TF13	?	?
Dowsby	TF13	pre- mid-18th century	?
Deeping Fen	TF12 & TF11	?	?
Bourne Fen	TF12	?	?
Cowbit	TF21	?	?
Fleet Fen	TF31	?	1793
Leake	TF35	?	?
Wrangle	TF45	?	?
Friskney	TF45	?	?
Wainfleet	TF45	?	?
Hagnaby	TF46	?	?

Sempringham Fen was a group of four decoys grouped closely together. Deeping Fen was a group of five decoys, three in TF12 and two in TF11. Wrangle was a group of three decoys. Friskney was a group of five decoys. Wainfleet had two decoys. The decoys in the list from Aslackby to Hagnaby were all working in the mid-18th century. One at Friskney was working up to 1878. Only three of them, two in Friskney and one in Wainfleet were still working after 1829 and only the one at Friskney after 1820. References: Wentworth-Day 1954, Lorand & Atkin 1989.

London (one site)

Site name	10-km square	Date built if known	Date abandoned if known
St James Park	TQ27	1665	?

Reference: Kear 1990.

Monmouthshire (two sites)

Site name	10-km square	Date built if known	Date abandoned if known
Nash	ST38	c.1810	1848–49
Wilcrick	ST38	c.1810	1848–49

Reference: Payne-Gallwey 1886.

Montgomeryshire (one site)

Site name	10-km square	Date built if known	Date abandoned if known
Lymore Park	SO29	?	?

Reference: Payne-Gallwey 1886.

Norfolk (31 sites)

Site name	10-km square	Date built if known	Date abandoned if known
Acle*	TG40?	see note below	1830s
Besthorpe	TM09	?	c.1815
Cawston*	TG12	?	?
Dersingham	TF63	1818	1870
Didlington Hall	TF08?	1855	early 20th century
Feltwell	TL79	?	?
Gunton Park	TG23	1803–04	?
Hemsby*	TG41	?	end 18th century
Hempstead*	TG 32/42	?	?
Hillington	TF72	?	?
Hilgay	TL69	?	1860
Hockwold	TL78	?	1838
Holkham	TF84	?	?
Langham	TG04	?	1854
Mautby *	TG41	?	1833

221

Site name	10-km square	Date built if known	Date abandoned if known
Merton	TL99	1886	still working 1890
Methwold	TL79	1806	1872
Micklemere	TL99	1830s	early 20th century
Rollesby*	TG41	?	?
Narford	TF71	1843	?
Northwold	TL79	?	still working 1930
Ranworth*	TG31	?	abandoned 1869
South Acre	TF71	1843	early 20th century
Sutton*	TG32	?	?
Stow Bardolph	TF60	?	1826
Waxham	TG42	c.1620	?
Westwick	TG22	?	early 20th century
Winterton	TG41/42	c.1807	?
Wolterton	TG13	?	?
Woodbastwick*	TG31	?	end 18th century
Wormegay	TF61	?	1838

Sir Thomas Browne, writing *c.*1663, commented on 'the very many decoys, especially between Norwich and the sea'. Those sites marked * all fit this description and may, therefore, be of 17th century date, although actual dates seem not to be recorded. References: Payne-Gallwey 1886, Stevenson & Southwell 1890, Southwell 1904, Riviere 1930, Wentworth-Day 1954.

Northamptonshire (two sites)

Site name	10-km square	Date built if known	Date abandoned if known
Borough Fen	TF10	1670	still used for ringing
Aldwincle	TL08	1885	?

References: Lilford 1895, Wentworth-Day 1954.

Nottinghamshire (five sites)

Site name	10-km square	Date built if known	Date abandoned if known
Wollaton Park	SK53	c.1825	1845
Haughton	SK67	ancient	still working 1907
Park Hall	SK56	?	still working 1907
Ossington Hall	SK76	c.1860	still working 1907
Annersley Park	SK55	?	still working 1907

The last four were all cage decoys. References: Payne-Gallwey 1886, Whitaker 1907.

Oxfordshire (one site)

Site name	10-km square	Date built if known	Date abandoned if known
Tythrop	SP70	probably 17th century	?

Reference: Payne-Gallwey 1886.

Pembrokeshire (one site)

Site name	10-km square	Date built if known	Date abandoned if known
Orielton	SR99	1868	1914–18, restored 1934, used to 1950

Reference: Payne-Gallwey 1886.

Shropshire (three sites)

Site name	10-km square	Date built if known	Date abandoned if known
Oakley Park	SO47	?	?
Sundorne Castle	SJ51	1780s	?
Aston Park	SJ32	before 1692	?

Reference: Payne-Gallwey 1886.

Somerset (14 or 15 sites)

Site name	10-km square	Date built if known	Date abandoned if known
Orions	ST34	before 1635	?
Sharpham Park	ST43	before 1635	?
Ivythorne, Sedgemoor	ST43	1825	still working 1886
Shapwick	ST43	1850	still working 1886
Meare	ST44	c.1807	1880s
Nyland	ST45	1678	?
Aller Moor	ST42	1676	1860s
Westbury	ST44	before 1635	?
Stoke	ST44	after 1802	short-lived
Compton Dundon	ST43	1695	?
Cheddar Water	ST45	?	?
Porlock	SS84	?	?

Ivythorne, Sedgemoor was a group of three Teal ponds. Payne-Gallwey mentions another decoy at Godney (ST44) but nothing was known about it save the name. References: Payne-Gallwey 1886, Southwell 1904, Whitaker 1918.

Staffordshire (one site)

Site name	10-km square	Date built if known	Date abandoned if known
Chillington	SO89	1825	1884

Reference: Payne-Gallwey 1886.

Suffolk (17 sites)

Site name	10-km square	Date built if known	Date abandoned if known
Fritton Lake*	TG40	?	in use 1932
Herringfleet*	TM49	end 17th century	c.1850
Hall Farm Fritton*	TG40	?	?
Nacton Wood/Orwell Park	TM23	1830–36	still working commercially 1932
Henham	TM47	?	?
Brantham	TM13	?	?
Camsey Ash	TM35	ancient	?
Chillesford	TM35	before 1807	1856
Euston Park	TL97	?	?
Flixton Lake	TM59	?	c.1828
Friston	TM45	?	c.1832
Iken	TM45	before 1807 and probably ancient	1914–1918
Lakenheath	TL68	?	1856
Nacton/Purdis Hall	TM24	18th century or earlier	1918
Worlingham	TM49	?	?
Benacre Broad	TM58	1880	never used

* All these 3 sites were round Fritton Lake and at least one of these sites may well have been in Norfolk. Ticehurst noted that there were originally seven pipes at the Fritton Decoy, declining to four, five declining to three at Herringfleet and one (only worked occasionally) at Hall Farm. He also noted that there were the remains of 13 other pipes, long disused, round the lake, making a maximum of 25 pipes in all. Nacton/Purdis Hall had two decoy ponds, one of which was for Teal. References: Ticehurst 1932; I am grateful to Dr Elizabeth Andrews for drawing my attention to the decoy at Henham.

Surrey (four sites)

Site name	10-km square	Date built if known	Date abandoned if known
Ottershaw Park	TQ06	?	end 18th century
Pyrford	TQ05	early 17th century	?
Clandon Park	TQ05	17th century	?
Virginia Water	SU96	?	1899

The decoy at Clandon Park was painted by Francis Barlow, who lived between 1626 and 1704. References: Bucknill 1900, Vandervell & Coles 1980.

Sussex (nine sites, ten if Peamarsh counted as two)

Site name	10-km square	Date built if known	Date abandoned if known
Tangmere	SU90	18th century or earlier	early 19th century
Tortington	TQ00	ancient	1840s
Poling	TQ00	17th century	1868
Glynde	TQ40	?	1780s
Firle Park	TQ40	?	c.1895
Ratton	TQ50	?	1840s
Glynleigh Manor Pevensey Levels	TQ60	before 1643	c.1880
Crowhurst	TQ71	?	c.1780s
Peasmarsh	TQ82	?	1820s

The Firle Park decoy was a house decoy with a single pipe. That at Peasmarsh was a double decoy with one pool for Teal. References: Payne-Gallwey 1886, Lennard 1905, Walpole-Bond 1938.

Warwickshire (three sites)

Site name	10-km square	Date built if known	Date abandoned if known
Stoneleigh	SP37	17th century	?
Coombe Abbey	SP37	1845	late 19th century
Packington Hall	SP28	1795	still working 1886

Coombe Abbey was a house decoy. References: Payne-Gallwey 1886, Heaton 2001.

Wiltshire (one site)

Site name	10-km square	Date built if known	Date abandoned if known
Hampworth	SU13	1882	?

A house decoy. Reference: Payne-Gallwey 1886.

Yorkshire (19 sites)

Site name	10-km square	Date built if known	Date abandoned if known
Balby Carr	SE50	1657	c.1778
Meaux[1]	TA03		
Watton[1]	TA05		
Scorborough[1]	TA04		
Holme[1]	SE83		
Sunk Island	TA21	end 17th century	never used
Escrick Park[2]	SE64	1830	1860
Osgodby	SE63	?	1878
Coatham[3]	NZ62	?	1872
Hornby Castle[4]	SE28	1854; 1882[4]	1885
Thirkleby[5]	SE47	1885[5]	?
Birdshall[6]	SE86	?	early 19th century
Thorne Waste[7]	SE17/27		

1. These four decoys were listed as ancient by Payne-Gallwey and Nelson and may have been constructed in the second half of the 17th century. They were all progressively abandoned with the drainage of the Holderness Carrs between 1762 and 1800. 2. Two decoys were constructed on this site but only one was in use at any time. 3. Nelson recorded a second decoy on the same estate but no record of it existed except the name Old Decoy. 4. The original decoy was dismantled in 1882 and moved to a new site. 5. Constructed by Sir Ralph Payne-Gallwey. This and Aldwincle in Northants, constructed by Lord Lilford, were the last decoys to be constructed in Britain. 6. A cage decoy. 7. Limbert recorded 6 decoys in this area of southeast Yorkshire, two near Goole, two near Thorne and two near Crowle, but the history of these sites is oddly difficult to disentangle. They were abandoned in the second half of the 19th century with extensive agricultural improvement. References: Payne-Gallwey 1886, Nelson 1907, Limbert 1978, 1982.

Appendix 3

Introductions and releases of Red-legged Partridge *Alectoris rufa* in Britain up to the end of the 19th century.

County	Place	Year	Reference
Berkshire	Windsor	1673	Gladstone 1930
Surrey	Wimbledon	1721–1729	Long 1981
Essex	St Osyth	*c.*1768	Fitter 1959
Suffolk	Sudbourne & Rendlesham	1770 or 1790	Ticehurst 1932
	Culford	1823	Ticehurst 1932
	Fourham	1824	Ticehurst 1932
	Cavenham	1824	Ticehurst 1932
Sussex	Harting	1776	Knox 1849
	Kirdford	1841	Borrer 1891
	Parham	First half of 19th century	Walpole-Bond 1938
Cornwall	St Austell	1808	Fitter 1959
Somerset	Cheddar	1817, unsuccessful	Somerset Ornithological Society 1988
Worcestershire	Witley	1820, unsuccessful	Harthan 1946
Norfolk	Aswelthorpe	1824	Stevenson 1866
	Many landowners	1820s	Stevenson 1866
Yorkshire	Bedale & Masham	1846–47	Nelson 1907
	Cleveland	1860	Nelson 1907
	Filey & Warter Priory	1892	Nelson 1907
	Pinchinthorpe	1890s	Nelson 1907
	West Riding	Before 1897	Nelson 1907
	Routon Grange	1906	Nelson 1907

County	Place	Year	Reference
Devon	Not recorded	*c.*1850	D'Urban & Mathew 1895
Rutland	Not recorded	1850	Haines 1907
Lancashire	Rufford	1854 & 1879	Oakes 1953
Hertfordshire	Theobalds	1855	Christy 1890
Hampshire	Beaulieu	1864	Kelsall & Munn 1905
	Heron Court	1867	Kelsall & Munn 1905
Oxfordshire	Chipping Norton	1868	Aplin 1889
Nottinghamshire	Clipstone	1872	Whitaker 1907
	Berry Hill	1880	Whitaker 1907
East Lothian	Gullane Hill	1876	Bolam 1912
Berwickshire	Edington	1887 & 1888	Bolam 1912
Dorset	Wimbourne & Poole	1888	Harting 1901
Shropshire	Not recorded	Late 19th century	Forrest 1899

Appendix 4

Ancient colonies (established in the 18th century or earlier) of the Black-headed Gull *Larus ridibundus* in England and Wales, by county.

Hampshire

Pewit Island in Portsmouth Harbour. Existed in early 17th century. It must then have been of considerable size, as Pennant recorded that it was worth £40 per year to the owner from the sale of young birds, a sum suggesting 1,000–1,600 young were taken annually. The colony probably died out in the 18th century, as there are no subsequent records. References: Pennant (1776), Clark & Eyre (1993).

Sussex

The Crumbles between Eastbourne and Pevensey. Certainly existed in the mid-17th century, when young gulls were sent from there to Lord Dacre at Herstmonceux Castle, costing 6d apiece. But there is no indication of the size of this colony, although Ticehurst suggested that young birds were sent to other local manors in season.

Winchelsea. Existed in the first half of the 17th century, and Ticehurst quoted evidence that the help of the Lord Warden of the Cinque Ports was sought to obtain recompense or recovery from a poaching foray in 1638, when young to the value of £5 or £6 were stolen (at the going rate of 6d each, some 200 birds or more). Ticehurst suggested, convincingly, that this must have been a large and valuable colony for the owner to seek such help in dealing with poachers.

Neither of these colonies is recorded as still extant in any early Sussex avifauna, and further breeding did not occur in the county until 1932. References: Lennard (1905), Ticehurst 1922, Walpole-Bond (1938).

Kent

Hoppen Pits, Dungeness. Apparently first documented in 1847, but Ticehurst noted that Black-headed Gulls had bred here 'from time immemorial'. Breeding in thousands in 1847. Exploitation for young birds not recorded, but large numbers of eggs taken, with records of up to three bushels being taken in a day and sent to London. Ticehurst remarked that 'if they were protected at all it was only for the greater benefit of the owner or tenant, but for the most part anyone was free to take their eggs for eating or selling and even the sailors from the Dutch pilot boats used to land and gather eggs by the basketful'. Such severe exploitation resulted in a steady decline and, by 1903, only 60 pairs remained. Reference: Ticehurst 1909.

Ticehurst apparently knew of no records of ancient colonies on the Kent side of the Thames estuary.

Essex

Pewit Island in Hamford Water. Certainly occupied in 1662, and bred there and latterly nearby on Horsey Island until the 1880s. Said to breed 'in great quantity' in the 17th century, and considerable numbers of young were harvested annually. Gurney (1921) records 10 dozen being taken in a day in July 1678. Eggs were harvested from Horsey Island by the people of Harwich.

Pewit Island in Blackwater river. Probably occupied in the 18th century, and certainly before 1833. 10,000 to 12,000 eggs taken annually in early 19th century, and sold at 4d per dozen. Deserted after 1887, but precise date unrecorded.

Foulness Island. A colony existed here in the 17th century but little was recorded about it, except that the gulls were abundant. References: Christy 1890, Gurney 1921. Christy also mentions old colonies, long defunct (see p. 147).

Suffolk

Alfred Newton told Stevenson & Southwell (1890) of an ancient colony at Brandon on the Suffolk/Norfolk border, which was extinct by *c.*1850, probably as the result of excessive egging in the 19th century.

Norfolk

Scoulton Mere near Hingham. An ancient colony perhaps in existence in the 13th or 14th centuries and occupied until 1964. Large numbers of young sent to London in the 17th century and probably much earlier, and eggs taken in large quantities in 19th century. Cott recorded maxima of 44,000 in 1840, 30,000 in 1845 and 16,000 in 1860, and an average of 5,775 a year from 1864 to 1919. Yarrell noted that a man and three boys found constant employment collecting them in May, sometimes gathering a thousand in a day. The eggs were sold on the spot at 4d a score and were regularly sent to the markets in Norwich and King's Lynn.

Stanford Warren. Existed before 1829 and probably earlier. On June 15th 1834, 'some thousands' were breeding. It was the practice here to collect the eggs up to the end of June, the eggs selling for 4d per dozen. The colony died out in around 1837.

Horsey. Enormous numbers bred in the 17th century, and young were sent in carts to Norwich and sold 'at small rates'. Country folk made much use of the eggs, which were also sent to Norwich. The colony died out with the enclosure and drainage of its site, and colonies established at Rollesby Broad and Hoveton in early 19th century probably stemmed from Horsey. The Hoveton gallery, which still exists, was managed for eggs, *c.* 2,000 being taken annually from the mid-19th century. References: Yarrell (1845), Stevenson & Southwell (1890), Gurney 1921, Riviere (1930), Cott (1953), Taylor *et al.* (1999).

Lincolnshire

Fens. Montagu (1833) noted plentiful birds in some of the Fens in the early 19th century, nesting in tufts of bog grass, and Pennant (1776) recorded them breeding there in the 18th century. Gough, in his edition of *Camden's Britannia* of 1806, described them as 'abounding'.

Twigmoor. A major colony in northwest Lincolnshire near Scawby. Colonised *c.*1840 from an older colony at Nathanland, which was drained *c.*1840. Birds from an old colony on

Manton Common, where they were much persecuted, probably also moved here. Reference: Blathwayt 1909.

Yorkshire

Thornton Bridge. A large colony with 'many thousands' breeding was recorded here in 1702 but it had disappeared by 1844.

Hornsea Mere. A large colony here was probably extant in the 17th century and was certainly occupied from before 1844 to the end of the 19th century.

Strensall Common, near York. Recorded by Yarrell & Saunders as an ancient site, but deserted in the late 19th century in spite of a brief attempt to recolonise it in the 1880s.

Thorne Waste. An ancient colony on this site, which Blathwayte recorded as near Crowle on the Lincolnshire border, was lost to drainage at an unrecorded date. Chislett noted that a number of other old colonies in Yorkshire were similarly lost.

Little exact data on actual numbers or the scale of exploitation seems to have been recorded for these sites, except that unrestrained egging was a serious problem in 19th century colonies, leading to the extermination of some. References: Yarrell & Saunders (1884), Nelson (1907), Blathwayte (1909), Chislett (1953), Cott (1953).

Northumberland

Pallinsburn. This colony, near Coldstream, existed certainly back to the mid-18th century, but Bolam gives no other details. He also referred to several other old breeding colonies, but without details. MacPherson noted that some of the young birds paid for in the Naworth accounts in the early 17th century probably came from Northumberland. References: MacPherson (1892), Bolam (1912).

Staffordshire

Norbury. A famous gullery on an island in a pool at this site was old in the mid-17th century, when visited by Willughby & Ray. It was clearly of considerable size and yielded 2200–3000 young birds annually. It died out in 1794. Ray (1678), McAldowie (1893).

Cheshire

Delamere Forest. A colony probably existed here in the early 17th century. Reference: Coward 1910.

Lancashire

What were probably ancient colonies existed on Walney Island, certainly from the 18th century, and where Fleetwood now stands; the latter was deserted in 1833 for Pilling Moss. But Mitchell gives no information on numbers or exploitation. Reference: Mitchell (1885).

Lakeland

Ravenglass. A very old site that was still occupied until the early 1980s, when it disappeared, possibly as the result of fox predation. Del Hoyo *et al.* noted that records of continuous, yearly egging went back to the 1600s and the colony had an annual yield of 30,000 eggs. It was protected from over-exploitation by careful local regulation. Young were also taken, if this was the main source for those recorded in the Naworth accounts, but MacPherson indicates that young were sent from a second site, noting that 'others were supplied from the colony of Black-headed Gulls near Muncaster by my Lady Savell'. References: MacPherson (1892), Del Hoyo *et al.* (1996) Lloyd *et al.* (1991).

Wales

Willughby & Ray visited Caldey, one of the Pembrokeshire islands, in June 1662 and recorded a substantial colony of Black-headed Gulls and terns. Nothing seems to have been recorded of it since. Pennant, in his *Tour of Wales*, recorded two colonies in North Wales, at Llyn Llydaw and Llyn Conwy about 1781. Both sites were deserted by the mid-19th century, if not before. References: Gurney 1921.

Appendix 5

Currency, weights and measures.

Values of money at different dates

The Bank of England gives the following historic values for £1.00 in 2011 (see www.bankofengland.co.uk/education):

    1850 – 1879 multiply by 110.45
    1880 multiply by 98.70
    1890 multiply by 105.43
    1900 multiply by 100.84
    1910 multiply by 96.64
    1920 multiply by 36.67
    1930 multiply by 53.63

Monetary values in the text are shown as pounds, shillings and pence as these are those recorded in earlier documents. They are written as £10–12s–6d, or as 7/6d for seven shillings and sixpence, or 10/- for ten shillings, and so forth. One modern penny (p) is worth 2.4 old pence (d), making one shilling worth 5p and one pound 100p. From the mid-19th century I have also shown today's equivalent value, as converted by the factors shown above.

Weights and measures

Weights and measures are given in the forms used in the documents quoted, with metric equivalents shown. One exception is the quarter, a measure of grain often quoted in early documents. A quarter was eight bushels, but the weight of bushels of wheat, barley and oats differ, wheat weighing 63 pounds (lbs) to the bushel, barley 56 and oats 42. Thus in no case is a 'quarter' a quarter of a ton, and the metric equivalents will differ.

Appendix 6

Scientific names of species mentioned in the text.
These names generally occur with the first appearance
of the species in each chapter; thereafter I have used just the
common name, to assist the flow of text.

Plants

Bird Cherry *Prunus padus*
Elder *Sambucus nigra*
Juniper *Juniperus communis*
Mistletoe *Viscum album*
Rowan *Sorbus aucuparia*
Service Tree *Sorbus domestica* or *torminalis*

Mammals

Bear (Brown Bear) *Ursus arctos*
Bison *Bison bonasus*
Elk *Alces alces*
Fallow Deer *Dama dama*
Hare (Brown Hare) *Lepus europaeus*
Otter *Lutra lutra*
Rabbit *Oryctolagus cuniculus*
Red Deer *Cervus elaphus*
Roedeer *Capreolus capreolus*
Wild Boar *Sus scrofa*

Birds

Arctic Tern *Sterna paradisaea*
Argus pheasants *Rheinardia ocellata,*
    Argusianus argus
Avocet (Pied Avocet) *Recurvirostra avosetta*
Baikal Teal *Anas formosa*
Barnacle Goose *Branta leucopsis*
Barn Owl *Tyto alba*
Barred Owl *Strix varia*
Bar-tailed Godwit *Limosa lapponica*
Bean Goose *Anser fabalis*
Bearded Tit *Panurus biarmicus*
Bee-eater (European Bee-eater) *Merops*
    apiaster

Bewick's Swan *Cygnus bewickii*
Bittern (Great Bittern) *Botaurus stellaris*
Blackbird (Common Blackbird) *Turdus*
    merula
Blackcap *Sylvia atricapilla*
Black Grouse *Tetrao tetrix*
Black-headed Gull *Larus ridibundus*
Black-necked Swan *Cygnus melanocoryphus*
Black-tailed Godwit *Limosa limosa*
Black Tern *Chlidonias niger*
Bluethroat *Luscinia svecica*
Boat-tailed Blackbird (Boat-tailed Grackle)
    Quiscalus major
Brambling *Fringilla montifringilla*
Brent Goose *Branta bernicla*
Brown Pelican *Pelecanus occidentalis*
Brünnich's Guillemot *Uria lomvia*
Bullfinch (Common Bullfinch) *Pyrrhula*
    pyrrhula
Calandra Lark *Melanocorypha calandra*
Canada Goose *Branta canadensis*
Canary *Serinus canaria*
Capercaillie (Western Capercaillie) *Tetrao*
    urogallus
Carolina Parakeet *Conuropsis carolinensis*
Cattle Egret *Bubulcus ibis*
Chaffinch *Fringilla coelebs*
Chough (Red-billed Chough) *Pyrrhocorax*
    pyrrhocorax
Cirl Bunting *Emberiza cirlus*
Common Gull (Mew Gull) *Larus canus*
Common Tern *Sterna hirundo*
Coot (Common Coot) *Fulica atra*
Cormorant (Great Cormorant) *Phalacrocorax*
    carbo

Corn Bunting *Emberiza calandra*

Corncrake (Landrail) *Crex crex*

Crane (Common Crane) *Grus grus*

Cream-coloured Courser *Cursorius cursor*

Crow (Carrion Crow) *Corvus corone*

Crossbill (Common Crossbill) *Loxia curvirostra*

Crowned Crane *Balearica pavonina*

Cuckoo (Common Cuckoo) *Cuculus canorus*

Curlew (Eurasian Curlew) *Numenius arquata*

Dartford Warbler *Sylvia undata*

Dotterel (Eurasian Dotterel) *Charadrius morinellus*

Double Crested Cormorant *Phalacrocorax auritus*

Dunlin *Calidris alpina*

Eagle Owl *Bubo bubo*

Eider (Common Eider) *Somateria mollissima*

Eskimo Curlew *Numenius borealis*

Falcated Duck (Falcated Teal) *Anas falcata*

Fieldfare *Turdus pilaris*

Firecrest *Regulus ignicapillus*

Flamingo (Greater Flamingo) *Phoenicopterus ruber*

Fulmar (Northern Fulmar) *Fulmarus glacialis*

Gannet (Northern Gannet) *Morus bassanus*

Garden Warbler *Sylvia borin*

Garganey *Anas querquedula*

Glossy Ibis *Plegadis falcinellus*

Goldcrest *Regulus regulus*

Golden Eagle *Aquila chrysaetos*

Golden Plover (European Golden Plover) *Pluvialis apricaria*

Goldeneye (Common Goldeneye) *Bucephala clangula*

Goldfinch *Carduelis carduelis*

Goosander *Mergus merganser*

Goshawk (Northern Goshawk) *Accipiter gentilis*

Grasshopper Warbler (Common Grasshopper Warbler) *Locustella naevia*

Gray Kingbird *Tyrannus dominicensis*

Great Auk *Pinguinus impennis*

Great Bustard *Otis tarda*

Great Crested Grebe *Podiceps cristatus*

Great Grey Shrike *Lanius excubitor*

Great Northern Diver *Gavia immer*

Great Reed Warbler *Acrocephalus arundinaceus*

Great Snipe *Gallinago media*

Great White Egret *Ardea alba*

Greater Bird-of-Paradise *Paradisaea apoda*

Greater White-fronted Goose *Anser albifrons*

Greenfinch (European Greenfinch) *Carduelis chloris*

Green Woodpecker *Picus viridis*

Greylag Goose *Anser anser*

Grey Heron *Ardea cinerea*

Grey Partridge *Perdix perdix*

Grey Phalarope *Phalaropus fulicarius*

Grey Plover *Pluvialis squatarola*

Guillemot (Common Guillemot) *Uria aalge*

Gull-billed Tern *Sterna nilotica*

Gyr Falcon *Falco rusticolus*

Hawfinch *Coccothraustes coccothraustes*

Hazel Grouse (Hazel Hen) *Bonasa bonasia*

Heath Hen *Tympanuchus cupido*

Herring Gull *Larus argentatus*

Hobby (Eurasian Hobby) *Falco subbuteo*

Hoopoe *Upupa epops*

House Sparrow *Passer domesticus*

Huia *Heteralocha acutirostris*

Impeyan Pheasant (Himalayan Pheasant) *Lophophorus impejanus*

Jackdaw (Western Jackdaw) *Corvus monedula*

Jack Snipe *Lymnocryptes minimus*

Jay (Eurasian Jay) *Garrulus glandarius*

Kentish Plover *Charadrius alexandrinus*

King Eider *Somateria spectabilis*

Kingfisher (Common Kingfisher) *Alcedo atthis*

Kittiwake (Black-legged Kittiwake) *Rissa tridactyla*

Knot (Red Knot) *Calidris canutus*

Labrador Duck *Camptorhynchus labradorius*

Lanner Falcon *Falco biarmicus*

Lapland Bunting *Calcarius lapponicus*

Lapwing (Northern Lapwing) *Vanellus vanellus*

Least Tern *Sterna antillarum*

Lesser Black-backed Gull *Larus fuscus*

Lesser Grey Shrike *Lanius minor*

Lesser Redpoll *Carduelis cabaret*

Linnet (Common Linnet) *Carduelis cannabina*

Little Bittern *Ixobrychus minutus*

Little Bustard *Tetrax tetrax*
Little Egret *Egretta garzetta*
Little Grebe (Dabchick) *Tachybaptus ruficollis*
Little Gull *Larus minutus*
Little Owl *Athene noctua*
Little Tern *Sterna albifrons*
Long-tailed Duck *Clangula hyemalis*
Mallard *Anas platyrhynchos*
Magpie (Black-billed Magpie) *Pica pica*
Manx Shearwater *Puffinus puffinus*
Marabou Stork *Leptoptilos crumeniferus*
Marsh Harrier (Eurasian Marsh Harrier)
    Circus aeruginosus
Marsh Warbler *Acrocephalus palustris*
Meadow Pipit *Anthus pratensis*
Merlin *Falco columbarius*
Mistle Thrush *Turdus viscivorus*
Monk Parakeet *Myiopsitta monachus*
Moorhen (Common Moorhen) *Gallinula
    chloropus*
Muscovy Duck *Cairina moschata*
Mute Swan *Cygnus olor*
Night Heron *Nycticorax nycticorax*
Nightingale (Common Nightingale) *Luscinia
    megarhynchos*
Nutcracker (Spotted Nutcracker) *Nucifraga
    caryocatactes*
Ortolan Bunting *Emberiza hortulana*
Osprey *Pandion haliaetus*
Ostrich *Struthio camelus*
Oystercatcher (Eurasian Oystercatcher)
    Haematopus ostralegus
Pallas's Warbler *Phylloscopus proregulus*
Passenger Pigeon *Ectopistes migratorius*
Peacock *Pavo cristatus*
Peregrine Falcon *Falco peregrinus*
Pheasant (Common Pheasant) *Phasianus
    colchicus*
Pied Flycatcher *Ficedula hypoleuca*
Pink-footed Goose *Anser brachyrhynchus*
Pine Grosbeak *Pinicola enucleator*
Pintail (Northern Pintail) *Anas acuta*
Pochard (Common Pochard) *Aythya ferina*
Pomarine Skua *Stercorarius pomarinus*
Ptarmigan (Rock Ptarmigan) *Lagopus muta*
Puffin (Atlantic Puffin) *Fratercula arctica*
Purple Gallinule *Porphyrio porphyrio*

Purple Heron *Ardea purpurea*
Quail (Common Quail) *Coturnix coturnix*
Raven (Common Raven) *Corvus corax*
Razorbill *Alca torda*
Red-backed Shrike *Lanius collurio*
Red Bird-of-Paradise *Paradisaea rubra*
Red-breasted Merganser *Mergus serrator*
Red Grouse *Lagopus lagopus*
Red Kite *Milvus milvus*
Red-legged Partridge *Alectoris rufa*
Red-necked Grebe *Podiceps grisegena*
Redshank (Common Redshank) *Tringa
    totanus*
Redwing *Turdus iliacus*
Richard's Pipit *Anthus richardi*
Ringed Plover *Charadrius hiaticula*
Ring-necked Parakeet (Rose-ringed Parakeet)
    Psittacula krameri
Robin (European Robin) *Erithacus rubecula*
Roller (Euopean Roller) *Coracias garrulus*
Rook *Corvus frugilegus*
Roseate Tern *Sterna dougallii*
Rose-coloured Starling (Rosy Starling) *Sturnus
    roseus*
Royal Tern *Sterna maxima*
Ruff *Philomachus pugnax*
Saker Falcon *Falco cherrug*
Sanderling *Calidris alba*
Sandwich Tern *Sterna sandvicensis*
Savi's Warbler *Locustella luscinioides*
Scarlet Ibis *Eudocimus ruber*
Scaup (Greater Scaup) *Aythya marila*
Scops Owl (Eurasian Scops Owl) *Otus scops*
Scoter (Common Scoter) *Melanitta nigra*
Serin *Serinus serinus*
Shag (European Shag) *Phalacrocorax aristotelis*
Shelduck (Common Shelduck) *Tadorna
    tadorna*
Shorelark (Horned Lark) *Eremophila alpestris*
Short-eared Owl *Asio flammeus*
Shoveler (Northern Shoveler) *Anas clypeata*
Siskin (Eurasian Siskin) *Carduelis spinus*
Skylark *Alauda arvensis*
Snipe (Common Snipe) *Gallinago gallinago*
Snow Bunting *Plectrophenax nivalis*
Snowy Egret *Egretta thula*
Song Thrush *Turdus philomelos*

Sooty Tern *Sterna fuscata*
Sparrowhawk (Eurasian Sparrowhawk) *Accipiter nisus*
Spoonbill (Eurasian Spoonbill) *Platalea leucorodia*
Spotted Redshank *Tringa erythropus*
Spotted Sandpiper *Actitis macularia*
Squacco Heron *Ardeola ralloides*
Starling (Common Starling) *Sturnus vulgaris*
Stock Dove *Columba oenas*
Stone Curlew *Burhinus oedicnemus*
Swallow (Barn Swallow) *Hirundo rustica*
Swan Goose *Anser cygnoides*
Tawny Owl *Strix aluco*
Teal (Eurasian Teal) *Anas crecca*
Tree Sparrow *Passer montanus*
Trumpeter Swan *Cygnus buccinator*
Tufted Puffin *Fratercula cirrhata*
Turkey (Wild Turkey) *Meleagris gallopavo*
Turnstone (Ruddy Turnstone) *Arenaria interpres*
Turtle Dove (European Turtle Dove) *Streptopelia turtur*
Twite *Carduelis flavirostris*
Velvet Scoter *Melanitta fusca*
Water Rail *Rallus aquaticus*

Waxwing (Bohemian Waxwing) *Bombycilla garrulus*
Wheatear (Northern Wheatear) *Oenanthe oenanthe*
Whimbrel *Numenius phaeopus*
Whistling Swan (Tundra Swan) *Cygnus columbianus*
White-breasted Kingfisher (Smyrna Kingfisher) *Halcyon smyrnensis*
White Stork *Ciconia ciconia*
White-tailed Eagle *Haliaeetus albicilla*
Whooper Swan *Cygnus cygnus*
Whooping Crane *Grus americana*
Wigeon (Eurasian Wigeon) *Anas penelope*
Wilson's Plover *Charadrius wilsonia*
Woodchat Shrike *Lanius senator*
Woodcock (Eurasian Woodcock) *Scolopax rusticola*
Woodlark *Lullula arborea*
Wood Pigeon (Common Woodpigeon) *Columba palumbus*
Wood Sandpiper *Tringa glareola*
Wood Warbler (Wood Wren) *Phylloscopus sibilatrix*
Wren (Winter Wren) *Troglodytes troglodytes*
Yellowhammer *Emberiza citrinella*

References

Adair, P. 1892. On the former abundance of the Quail (*Coturnix communis* Bonnaterre) in Wigtownshire. *Annals of Scottish Natural History* 1: 168–171.

Alexander, B. 1896. Ornithological notes from Romney Marsh. *Zoologist* 1896: 246–253.

Allen, D. E. 1976. *The Naturalist in Britain*. Penguin. Harmondsworth.

Almond, R. 2003. *Medieval Hunting*. Sutton Publishing. Stroud.

Aplin, O. V. 1889. *The Birds of Oxfordshire*. Oxford University Press. Oxford.

Armstrong, E. A. 1979. The Crane in the British Isles and Crane traditions as evidence of culture diffusion. In Lewis, J. C. 1979, *Proceedings 1978 Crane Workshop*. Colorado State University Printing Service.

Aubrey, J. 1969. *Natural History of Wiltshire*. David & Charles reprint of a 17th publication. Newton Abbot.

Babington, Rev. C. 1884–86. *Catalogue of the Birds of Suffolk*. Van Voorst. London.

Bailey, M. 2007. *Medieval Suffolk: an economic and social history, 1200–1500*. Boydell Press. Woodbridge.

Baines, D. & Hudson, P. J. 1995. The decline of Black Grouse in Scotland and northern England. *Bird Study* 42: 122–131.

Baldwin, J. R. 1974. Seabird fowling in Scotland and Faroe. *Folklife* 12: 60–130.

Ballance, D. K. 2006. *A History of the Birds of Somerset*. Isabelline Books. Penrhyn.

Bannerman, D. A. 1963. *The Birds of the Atlantic Islands* vol.1. Oliver & Boyd. London

Barnes, J. 1997. *The Birds of Caernarfonshire*. Cambrian Ornithological Society.

Baxter, E. V. & Rintoul, L. J. 1953. *The Birds of Scotland*. Oliver & Boyd. Edinburgh.

Benson, C. E. 1922. Egret farming in Sind. *Journal of the Bombay Natural History Society* 28: 748–750.

Bent, A. C. 1919. *Life Histories of North American Diving Birds*. Bulletin 107. United States National Museum

Bent, A. C. 1921. *Life Histories of North American Gulls and Terns*. Bulletin 113. United States National Museum.

Bent, A. C. 1923 & 1925. *Life Histories of North American Wild Fowl*. Bulletin 130. United States National Museum.

Bernis, F. 1960. About wintering and migration of the Common Crane (*Grus grus*) in Spain. *Proceedings of the 12th International Ornithological Congress*. Helsinki. pp. 110–117.

Bewick, T. 1826. *History of British Birds*. 6th edition. Bewick. Newcastle.

Bijleveld, M. 1974. *Birds of Prey in Europe*. Macmillan. London.

Birch, G. 1921. Egret farming in Sind. *Journal of the Bombay Natural History Society* 27: 944–947.

Bircham, P. 1989. *The Birds of Cambridgeshire*. Cambridge University Press.

Bircham, P. 2007. *A History of Ornithology*. Collins. London.

Birdlife International. 2004. *Birds in Europe: population estimates, trends and conservation status*. Cambridge UK. Birdlife International. (Birdlife Conservation Series No.12).

Blanning, T. 2007. *The Pursuit of Glory: Europe 1648–1815*. Penguin Allen Lane. London.

Blathwayt, Rev. F. L. 1909. Lincolnshire Gulleries. *Zoologist* 1909: 139–144.

Boisseau, S. & Yalden, D. W. 1998. The former status of the Crane *Grus grus* in Britain. *Ibis* 140: 482–500.

Bolam, G. 1912. *The Birds of Northumberland and the Eastern Borders*. Blair. Alnwick.

Booth, E. T. 1878. The Migration of Birds in Autumn. *Zoologist* 1878: 100–102.

Booth, E. T. 1881–87. *Rough Notes on the Birds Observed during 25 years shooting and collecting in the British Islands*. R. H. Porter. London.

Boothroyd, G. 1985. *The Shotgun: History and Development*. A&C Black. London (1991 edition published by Safari Press, California).

Borrer, W. 1891. *The Birds of Sussex*. R. H. Porter. London.

Bourne, W. R. P. 1981. The birds and animals consumed when Henry VIII entertained the King of France and the Count of Flanders at Calais in 1532. *Archives of Natural History* 10: 331–333.

Bourne, W. R. P. 1999a. The past status of the herons in Britain. *Bulletin British Ornithologists Club* 119(3): 192–196.

Bourne, W. R. P. 1999b. Information in the Lisle

letters from Calais in the early 16th century relating to the development of the English bird trade. *Archives of Natural History* 26: 349–368.

Bourne, W. R. P. 2003. Fred Stubbs, Egrets, Brewes and climatic change. *British Birds* 96: 332–339.

Bourne, W. R. P. 2006. Birds eaten by King James V of Scotland in 1525–33 and the prices decreed by his daughter Mary Queen of Scots in 1551. *Archives of Natural History* 33: 135–139.

Bourne, W. R. P. 2007. The early history of Scottish birds. In Forrester, R. W. & Andrews, I. J. (eds). 2007. *The Birds of Scotland*. Scottish Ornithologists' Club. Aberlady.

Boyce, W. 1877. Migration of Birds on the Yorkshire coast. *Zoologist* 1877: 42–43.

Brander, M. 1971. *Hunting & Shooting: from earliest times to the present day*. G. P. Putnam's Sons. New York.

Brentnall, H. C. (ed.) 1947. A Longford Manuscript. *The Wiltshire Magazine* 52: 1–56.

Brown, A. & Grice, P. 2005. *Birds in England*. T & AD Poyser. London.

Brusewitz, G. 1969. *HUNTING: Hunters, game, weapons and hunting methods from the remote past to the present day*. George Allen & Unwin. London.

Buckley, T. E. 1892. Contributions to the Vertebrate Fauna of Sutherland and Caithness. *Annals of Scottish Natural History*. 1892: 156–167.

Bucknill, J. A. S. 1900. *The Birds of Surrey*. R. H. Porter. London.

Bull, H. G. 1888. *Notes on the Birds of Herefordshire*. Jakeman & Carver. London.

Burke, T. 1940. *The streets of London through the centuries*. Batsford, London.

Burton, J. F. 1956. Report on the National Census of Heronries 1954. *Bird Study* 3: 42–73.

Burton, J. F. 1995. *Birds and Climate Change*. Christopher Helm. London.

Byrne, M. St C. & Boland, B. (eds). 1983. *The Lisle Letters: an abridgement*. University of Chicago Press. Chicago.

Cade, T. J. 1982. *The Falcons of the World*. Collins. London.

Cantor, L. M. 1982a. Forests, Chases, Parks and Warrens. In Cantor, L. M. 1982 (ed.). *The English Mediaeval Landscape*. Croom Helm. London.

Cantor, L. M. 1982b. Introduction. In Cantor, L. M. 1982 (ed). *The English Mediaeval Landscape*. Croom Helm. London.

Chalmers-Hunt, J. M. 1976. *Natural History Auctions 1700–1972: A Register of Sales in the British Isles*. Sotheby Parke Bernet. London.

Chapman, A. 1897. *Wild Norway*. Edward Arnold. London.

Chapman, A. & Buck, W. J. 1893. *Wild Spain*. Gurney & Jackson. London.

Chapman, F. M. 1943. *Birds and Man*. American Museum of Natural History Guide Leaflet series No.115. New York.

Chevenix Trench, C. C. 1922. Egret farming in India. *Journal of the Bombay Natural History Society* 28: 751–752.

Chislett, R. 1953. *Yorkshire Birds*. A. Brown & Sons. London.

Christy, R. M. 1890. *The Birds of Essex*. Simpkin, Marshall, Hamilton, Kent & Co. London.

Christy, R. M. 1903. *Birds in the Victoria History of the County of Essex*. London.

Chute, R. 2002. *Shooting sitting*. Foxbury Press. Winchester.

Claessens, O. 1992. La situation du Bruant Ortolan *Emberiza hortulana* en France et en Europe. *Alauda* 60: 65–76.

Clark, G. 1948. Fowling in Prehistoric Europe. *Antiquity* 22:116–130.

Clark, J. M. & Eyre, J.A. (eds.) 1993. *Birds of Hampshire*. Hampshire Ornithological Society.

Clark Kennedy, A. W. M. 1868. *The Birds of Berkshire and Buckinghamshire*. Ingalton & Drake. Eton.

Cobbett, W. *Rural Rides*. Everyman's Library Edition 1957. J. M. Dent & Sons. London.

Cocker, M. & Mabey, R. 2005. *Birds Britannica*. Chatto & Windus. London.

Cohen, E. 1963. *The Birds of Hampshire and the Isle of Wight*. Oliver & Boyd. Edinburgh.

Cole, A. C. 2006. *The egg dealers of Great Britain*. Peregrine Books. Leeds.

Cole, A. C. & Trobe, W. M. 2000. *The Egg Collectors of Great Britain and Ireland*. Peregrine Books. Leeds.

Cole, A. C. & Trobe, W. M. 2011. *The Egg Collectors of Great Britain and Ireland: an update*. Peregrine Books. Leeds.

Congreve, W. M. & Freme, S. W. P. 1930. Seven weeks in Eastern and Northern Iceland. *Ibis* 12th series; VI: 197–228.

Cook, W. A. 1960. The numbers of duck caught in Borough Fen Decoy 1776–1959. *Wildfowl* 11: 118–122.

Cooper, W. D. 1863. Produce of and Supplies from Sussex. *Sussex Archaeological Collections* 17: 115–122.

Cordeaux, J. 1864. Remarks on the Birds seen during a visit to Flamborough, in the last fortnight of July 1864. *Zoologist* 1864: 9243–9247.

Cordeaux, J. 1872. *Birds of the Humber District*. Van Voorst. London.

Cordeaux, J. 1885. Ornithological notes from the

east coast in the spring of 1885. *The Naturalist* 10: 267–269.

Cott, H. B. 1953, 1954. The exploitation of wild birds for their eggs. *Ibis* 95: 409–449 & 643–675 and 96: 129–149.

Coward, T. A. 1910. *The Vertebrate Fauna of Cheshire and Liverpool Bay*. Witherby. London

Coward, T. A. 1922. *Bird Haunts and Nature Memories*. Warne. London.

Coward, T. A. & Oldham, C. 1900. *The Birds of Cheshire*. Sherratt & Hughes. Manchester.

Cramp, S. & Simmons, K. E. L. 1977–1979. *The Birds of the Western Palaearctic* vols. I-II. Oxford University Press. Oxford.

Cummins, J. 1988. *The Hound and the Hawk: the art of mediaeval hunting*. Phoenix Press. London.

Darby, H.C. 1934. Notes on the birds of the undrained Fen. In Lack. D. *The Birds of Cambridgeshire*. Cambridge Bird Club. Cambridge.

Darby, H.C. 1974. *The Mediaeval Fenland*. 2nd edition. David & Charles. Newton Abbot.

Darby, W. J., Ghalioungui, P & Grivetti, L. 1977. *Food: the Gift of Osiris*. Academic Press. London.

Darling, F. F. & Morton Boyd, J. 1964. *The Highlands and Islands*. Collins. London

Dawnay, A. 1972. Exploitation. In Scott, P. & the Wildfowl Trust. 1972. *The Swans*. Michael Joseph. London.

Deby, J. 1846. Notes on the Birds of Belgium. *Zoologist* 1846: 1251–1261.

Delaney, S. & Scott, D. 2002. *Waterbird population estimates*. Wetlands International Global Series No.12.

Del Hoyo, J., Elliot, A. & Christie, D. A. (eds). 1992, 1996, 1999, 2009. *Handbook of the Birds of the World: vols 1, 3, 5 & 14*. Lynx Edicions. Barcelona.

Dementiev, G. P., Gladkov, N. A., Isakov, Yu. A., Kartashev, N. N., Kirikov, S. V., Mikheev, A. V. & Ptushenko, E. S. 1967. *Birds of the Soviet Union*. Israel Programme for Scientific Translations. Jerusalem.

Diamond, J. 2005. *Collapse: how societies choose to fail or survive*. Penguin Books. London.

Dobbs, A. (ed.) 1975. *The Birds of Nottinghamshire*. David & Charles. Newton Abbot.

Donald, P. F. 2004. *The Skylark*. T & AD Poyser. London.

Doughty, R. W. 1975. *Feather Fashions and Bird Preservation: a Study in Nature Protection*. University of California Press. Berkeley, Los Angeles & London.

Doughty, R. W. 1979. Eider husbandry in the North Atlantic: trends and prospects. *Polar Record* 19: 447–459.

D'Urban, W. S. M. & Mathew, Rev. M. A. 1895. *The Birds of Devon*. R. H. Porter. London.

Durnford, H. 1874. Ornithological Notes on the North Frisian Islands and adjacent Coast. *Ibis* (3) 4: 391–406.

Dutton, J. 1867. Guillemot and Razorbill near Eastbourne. *Zoologist* May 1867 759–760.

Ellis, E. A. 1965. *The Broads*. Collins. London.

Ehrlich, P. R., Dobkin, D. S. & Wheye, D. 1994. *The Birdwatchers Handbook*. Oxford University Press. Oxford.

Ennion, E. A. R. 1949. *Adventurers Fen*. Herbert Jenkins. London.

Evans, A. H. 1903. *Turner on birds*. Cambridge University Press. Cambridge.

Evans, A. H. 1911. *A Vertebrate Fauna of the Tweed Area*. David Douglas. Edinburgh.

Fadden, K. J. 2011. Location and Review of an ancient Duck Decoy in the Parish of Houghton Conquest. *Ampthill & District Archaeological and Local History Society.*

Fagan, B. 2000. *The Little Ice Age: how climate made history*. Basic Books. New York.

Fielden, H. W. 1870. Increase of Rock Birds at Flamborough. *Zoologist* 1870: 2262.

Fielden, H. W. 1872. The Birds of the Faroe Islands. *Zoologist* 1872: 3277–3294.

Fisher, C. T. 1997. Past human exploitation of birds on the Isle of Man. *International Journal of Osteoarchaeology* 7: 292–297.

Fisher, J. 1952. *The Fulmar*. Collins. London.

Fisher, J. 1966. *The Shell Bird Book*. Ebury Press and Michael Joseph. London.

Fisher, J. & Lockley, R.M. 1954. *Sea-birds*. Collins. London.

Fisher, J. & Vevers, H. G. 1943. The breeding distribution, history and population of the North Atlantic Gannet (*Sula bassana*). *Journal of Animal Ecology* 12: 173–213

Fisher, J. & Waterston, G. 1941. The breeding distribution, history and population of the Fulmar (*Fulmaris glacialis*) in the British Isles. *Journal of Animal Ecology* 10: 204–272.

Fitter, R. S. R. 1959. *The Ark in our midst*. Collins. London.

Flint, V. E., Boehme, R. L., Kostin, Y. W. & Kuznetsov, A. A. 1984. *A Field Guide to the Birds of the USSR*. Princeton University Press.

Forrest, H. E. 1899. *The Fauna of Shropshire*. Terry & Co. London.

Forrest, H. E. 1907. *The Fauna of North Wales*. Witherby. London.

Fray, R., Davis, R., Gamble, D,. Harrop, A. & Lister, S. 2009. *The Birds of Leicestershire and Rutland*. Christopher Helm. London.

Frost, C. 1987. *A History of British Taxidermy*. Privately published. Long Melford.

Furbank, P. N., Owens W. R. & Coulson, A. J. (eds). 1991. *Daniel Defoe: A Tour Through the Whole Island of Great Britain*. Yale University Press. London.

Gatke, H. 1895. *Heligoland: An ornithological observatory*. David Douglas. Edinburgh.

Gibbons, D & Dudley, S. 1993. Quail. In Gibbons, D. W., Reid, J. B. & Chapman R. A. 1993. *The New Atlas of Breeding Birds in Britain and Ireland: 1988–1991*. T & AD Poyser. London.

Gibson, E. 1920. Further ornithological notes from the neighbourhood of Cape San Antonio, Buenos Ayres. *Ibis* 11: 1–97.

Gilbert, J. M. 1979. *Hunting and Hunting Reserves in Mediaeval Scotland*. John Donald. Edinburgh.

Gladstone, H. 1943. Comparative prices of game and wildfowl in 1512, 1757, 1807, 1922, 1941 and 1942. *British Birds* 36:122–131.

Gladstone, H. S. 1924. The distribution of Black Grouse in Great Britain. *British Birds* 18: 66–68.

Gladstone, H. S. 1930. *Record Bags and Shooting Records*. 2nd revised edition. Facsimile reprint by Signet Press 1995.

Glegg, W. E. 1929. *A History of the Birds of Essex*. Witherby. London.

Glegg, W. E. 1935. *A History of the Birds of Middlesex*. Witherby. London.

Glegg, W. E. 1943, 1944. The Duck decoys of Essex. *Essex Naturalist*. 27: 191–207 & 211–225.

Glutz von Blotzheim, U. N., Bauer, K. M. & Bezzel, E. 1973. *Handbuch der Vogel Mitteleuropas vol 5*. Akademische Verlagsgesellschaft AULA-Verlag. Weisbaden.

Goodman, S. M. & Meininger, P. L. (eds). 1989. *The Birds of Egypt*. Oxford University Press. Oxford.

Graham, H. D. 1890. *The Birds of Iona and Mull 1852–70*. David Douglas. Edinburgh.

Gray, R. 1871. *The Birds of the West of Scotland*. Murray. Glasgow.

Gray, R. & Hussey, A. 1860. The trade in Goldfinches. *Zoologist* 18: 711–714.

Green, G. 2004. *The Birds of Dorset*. Christopher Helm. London.

Greenoak, F. (ed). 1986. *The Journals of Gilbert White. Vol:1*. Century. London.

Greenway, J. C. 1967 (2nd edition). *Extinct and Vanishing Birds of the World*. Dover Publications, Inc. New York.

Greenwood, J. J. D. 2012. John Nelder: statistics, birdwatching and the Hastings Rarities. *British Birds* 105: 733–737.

Griffith, J. E. 2000. Iceland. Part 4. *Bulletin of the Jourdain Society* 14: no: 13.

Grinnell, G. B. 1901. *American Duck Shooting*. 1991 reprint by Stackpole Books. Harrisburg.

Guest, J. P., Elphick, D., Hunter, J. S. A. & Norman, D. 1992. *Breeding Bird Atlas of Cheshire and Wirral*. Cheshire and Wirral Ornithological Society. Chester.

Gurney, D. 1834. Extracts from the Household and Privy Purse Accounts of the Lestranges of Hunstanton from AD 1519 to AD 1578. *Archaeologia* 25: 411–569.

Gurney, J. H. 1870. Leadenhall Market. *Zoologist* Dec.1870: 2393–2394.

Gurney, J. H. 1899. The Bearded Titmouse (*Panurus biarmicus Lin.*). *Transactions of the Norfolk & Norwich Naturalists Society* VI: 429–438.

Gurney, J. H. 1913. *The Gannet: a bird with a history*. Witherby. London.

Gurney, J. H. Jr. 1883. Imported Game Birds in English Markets. *Zoologist* series 3, vol. 7: 300–301.

Gurney, J. H. Jr. 1921. *Early Annals of Ornithology*. Witherby. London.

Hagemeijer, W. J. M. & Blair, M. J. (eds). 1997. *The EBCC Atlas of European Breeding Birds: their distribution and abundance*. T & AD Poyser. London.

Haines, C. R. 1907. *Notes on the Birds of Rutland*. R. H. Porter. London.

Hall, J. J. 1982. The Cock of the Wood. *Irish Birds* 2: 38–47.

Hansen, K. 2002. *A Farewell to Greenland's wildlife*. Gads Forlag. Copenhagen.

Hardy, R. 1992 (3rd ed.). *Longbow: a social and military history*. Patrick Stephens Ltd. Sparkford.

Harradine, J. 1986. Jack Snipe. In Lack, P. (ed.) *The Atlas of Wintering Birds in Britain and Ireland*. T & AD Poyser. Calton.

Harrison, C. & Reid-Henry, D. 1988. *The History of the Birds of Britain*. Collins. London.

Harrison, G. R., Dean, A. R., Richards, A. J. & Smallshire, D. *The Birds of the West Midlands*. West Midlands Bird Club. Studley.

Harrison, G. & Harrison, J. 2005. *The New Birds of the West Midlands*. West Midlands Bird Club.

Harrison, T. H. & Hollom, P. A. D. 1932. The Great Crested Grebe Enquiry 1931. *British Birds* 26: 62–92,102–131,142–155, 174–195.

Harrison, T. P. & Hoeniger, F. D. (eds). 1972. *The Fowles of Heauen or History of Birdes by Edward Topsell*. University of Texas. Austin.

Harrop, A. H. J., Collinson, J. M. & Melling, T. 2012. What the eye doesn't see: the prevalence of fraud in ornithology. *British Birds* 105: 236–257.

Hartert, E. & Jourdain, F. C. R. 1920. *The Birds of Buckinghamshire and the Tring Reservoirs*. Reprinted from *Novitates Zoologicae* vol. XXVII.

Harthan, A. J. 1946. *The Birds of Worcestershire*. Littlebury. Worcester.

Harting, J. E. 1866. *The Birds of Middlesex*. Van Voorst. London.

Harting, J. E. 1871. *The Birds of Shakespeare*. Van Voorst. London.

Harting, J. E. 1872. British Heronries. *Zoologist* 1872: 3262.

Harting, J. E. 1877. On the former nesting of the Spoonbill in the County of Sussex. *Zoologist* 1877: 425–429.

Harting, J. E. 1879. Early mention of the Hoopoe as a British bird. *Zoologist* no: 337–338.

Harting, J. E. 1880. Some notes on Hawking, as formerly practised in Norfolk. *Transactions of the Norfolk and Norwich Naturalists Society* I: 79–94.

Harting, J. E. 1882. The Crane. *The Field*, December 23rd.

Harting J. E. 1886. Some further notes on Hawking in Norfolk. *Transactions of the Norfolk and Norwich Naturalists Society* IV: 248–254

Harting, J. E. 1890. Of Hawks and Hawking in Essex in olden time. In Christy, R.M. 1890. *The Birds of Essex*. Simpkin, Marshall, Hamilton, Kent & Co. London.

Harting, J. E. 1901. *A Handbook of British Birds*. 2nd edition. J.C. Nimmo. London.

Harvey, P. D. A. 1991. Farming Practice and Techniques in the Home Counties. In Miller, E. (ed). *The Agrarian History of England and Wales*. Vol. III. Cambridge University Press. Cambridge.

Harvie-Brown, J. A. 1906. *A Fauna of the Tay Basin and Strathmore*. David Douglas. Edinburgh.

Harvie-Brown, J. A. & Buckley, T. E. 1887. *A Vertebrate Fauna of Sutherland, Caithness and West Cromarty*. David Douglas. Edinburgh.

Harvie-Brown, J. A. & Buckley, T. E. 1892. *A Vertebrate Fauna of Argyll and the Inner Hebrides*. David Douglas. Edinburgh.

Harvie-Brown, J. A. & Buckley, T. E. 1895. *A Fauna of the Moray Basin*. David Douglas Edinburgh.

Hastings, M. 1981. *The Shotgun*. David & Charles. Newton Abbot.

Haverschmidt, F. 1943. De Goudpluvierenvangst in Nederland [with English summary]. *Ardea* 32: 35–74.

Haverschmidt, F. 1963. *The Black-tailed Godwit*. Brill. Leiden.

Hawker, P. (1893). *The Diary of Colonel Peter Hawker 1802–1853*. Longman, Green & Co. London.

Heathcote, A., Griffin, D. & Salmon, H. M. (eds). 1967. *The Birds of Glamorgan*. Cardiff Naturalists Society.

Heaton, A. 2001. *Duck Decoys*. Shire Publications. Princes Risborough.

Hele, N. F. 1870. *Notes or Jottings about Aldeburgh Suffolk*. John Russel Smith. London

Herriott, S. 1968. *A Directory of British Taxidermists*. Leicester Museum. Leicester.

Hill, D. & Robertson, P. 1988. *The Pheasant: Ecology, management and conservation*. Blackwell. Oxford.

Hobusch, E. 1980. *Fair Game: a history of hunting, shooting and animal conservation*. Arco Publishing, Inc. New York.

Holloway, S. 1996. *The Historical Atlas of Breeding Birds in Britain and Ireland, 1875–1900*. T & AD Poyser. London.

Hope, A. 1990. *Londoners' Larder: English cuisine from Chaucer to the Present*. Mainstream Publishing. Edinburgh.

Houlihan, P. F & Goodman, S. M. 1986. *The Birds of Ancient Egypt*. Aris & Phillips. Warminster.

Hudson, W. H. 1900. *Nature in Downland*. Longman Green & Co. London.

Hume, J. P. & Walters, M. 2012. *Extinct Birds*. T & AD Poyser. London.

Hywel, W. 1973. *Modest millionaire: A biography of Vivian Hewitt*. Gwasg Gee. Denbigh.

Ingram, G. C. S. & Salmon, H. M. 1954. *A Handlist of the Birds of Carmarthenshire*. West Wales Naturalists' Trust. Haverfordwest.

Ingram, G. C. S. & Salmon, H. M. 1957. The Birds of Brecknock. *Brycheiniog* 3: 182–259.

Inskipp, T. & Gammell, A. 1979. The extent of world trade in birds and the mortality involved. *Bulletin of the International Council for Bird Preservation* 13: 98–103.

Israel, J. I. 1995. *The Dutch Republic: its Rise, Greatness, and Fall 1477–1806*. Oxford University Press. Oxford.

James, A. 1986. *Memoirs of a Fen Tiger*. David & Charles. Newton Abbot.

Johnsgard, P. A. 1983. *Cranes of the World*. Croom Helm. London.

Jones, P. E. 1965. *The Worshipful Company of Poulters of the City of London (2 edition.)*. Oxford University Press. Oxford.

Jourdain, F. C. R. 1922. On the Birds of Spitsbergen and Bear Island. *Ibis* (11) 4: 159–179.

Kear, J. 1990. *Man and Wildfowl*. T & AD Poyser. London.

Kearton, R. 1909. *With Nature and a Camera*. Cassell and Company. London. First published 1897.

Kelsall, J. E. & Munn, P. W. 1905. *The Birds of Hampshire and the Isle of Wight*. Witherby. London.

Kennedy, P. G., Ruttledge, R. F. & Scroope, C. F. 1954. *The Birds of Ireland*. Oliver & Boyd. Edinburgh and London.

Key, H. A. S. 1955. The Tempsford Duck Decoy. *The Bedfordshire Naturalist for 1954*: 24–28.

Knapp, J. L. 1829. *The Journal of a Naturalist*. John Murray. London.

Knox, A. E. 1849. *Ornithological Rambles in Sussex*. Van Voorst. London.

Knox, A. E. 1850. *Game Birds and Wildfowl: their friends and foes*. Van Voorst. London.

Kortright, F. H. 1942. *Swans, Geese and Ducks of North America*. Stackpole, Harrisburg and Wildlife Management Institute, Washington.

Knystautas, A. 1987. *The Natural History of the USSR*. Century Hutchinson. London.

Lack, D. 1934. *The Birds of Cambridgeshire*. Cambridge Bird Club. Cambridge.

Lamb, H. H. 1995. *Climate History and the Modern World (2nd edition)*. Routledge. London.

Lamb, T. 1880 (but written in 1814). Ornithologia Bercheria. *Zoologist* 1880: 313–325.

Landsborough Thompson, Sir A. 1964. *A New Dictionary of Birds*. Nelson. London.

Lennard, T.B. 1905. Extracts from the Household Account Book of Herstmonceaux Castle, August 1643 to December 1649. *Sussex Archaeological Collections* 48: 104–137.

Le Strange, H. 1920. A Roll of Household Accounts of Sir Hamon Le Strange of Hunstanton, Norfolk 1347–8. *Archaeologia* 69: 111–120.

Lever, C. 2005. *Naturalised Birds of the World*. T & AD Poyser. London.

Lewis, S. 1936. Birds of the Island of Steep Holm. *British Birds* 30: 219–223.

Lewis, S. 1952. *The Breeding Birds of Somerset and their Eggs*. Privately published.

Lilford, Lord. 1895. *Notes on the Birds of Northamptonshire and Neighbourhood*. R. H. Porter. London.

Limbert, M. 1978. The old duck decoys of south-east Yorkshire. *The Naturalist* 103: 95–103.

Limbert, M. 1982. The duck decoys of south-east Yorkshire: an addendum. *The Naturalist* 107: 69–71.

Lippens, L. & Wille, H. 1969. Le Heron Bihoreau *Nycticorax n. nycticorax* (Linne 1758) en Belgique et en Europe Occidentale. *Gerfaut* 59/2: 124–156.

Lloyd, C., Tasker, M.L. & Partridge, K. 1991. *The Status of Seabirds in Britain and Ireland*. T & AD Poyser. London.

Lockley, R. M. 1953. *Puffins*. Dent & Sons. London

Long, J. L. 1981. *Introduced Birds of the World*. David & Charles. Newton Abbot.

Lorand, S. & Atkin, K. 1989. *The Birds of Lincolnshire and South Humberside*. Leading Edge. Burtersett.

Lovegrove, R. 2007. *Silent Fields: the long decline of a nation's wildlife*. Oxford University Press. Oxford.

Lovegrove, R., Williams, G. & Williams, I. 1994. *Birds in Wales*. T & AD Poyser. London.

Lowe, F. A. 1954. *The Heron*. Collins. London.

Loyd, L. R. W. 1929. *The Birds of South East Devon*. Witherby. London.

Lubbock, Rev. R. 1845. *Observations on the Fauna of Norfolk*. Longmans. London.

Macgillivray, W. 1837–1852. *History of British Birds*. Scott, Webster & Geary. London.

Macgregor, A. 1989. Animals and the early Stuarts: hunting and hawking at the court of James 1 and Charles 1. *Archives of Natural History* 16: 305–318.

MacKenzie, Rev. N. 1905. Notes on the Birds of St Kilda. *Annals of Scottish Natural History* 14: 75–80, 141–53.

Mackenzie, O. 1949. *A Hundred Years in the Highlands*. First published 1921, new revised edition 1949. G. Bles. London.

Macpherson, H. A. 1892. *A Vertebrate Fauna of Lakeland*. David Douglas. Edinburgh

Macpherson, H. A. 1897. *History of Fowling*. David Douglas. Edinburgh.

Madsen, J., Cracknell, G. & Fox, A. eds. 1999. *Goose populations of the Western Paleaearctic*. Wetlands International Publication No. 48. National Environment Research Institute. Denmark.

Magnin, G. 1987. An account of the illegal catching and shooting of birds in Cyprus during 1986. *ICBP Study Report No. 21*. ICBP.

Mansell-Pleydell, J. C. 1888. *The Birds of Dorsetshire*. R. H. Porter. London.

Marchant, J. 2004.. Monitoring UK Herons. *BTO News* 250: 22–23.

Marchant, J. V. R. & Watkins, W. 1897. *Wild Birds Protection Acts*. R.H. Porter. London.

Marchington, J. 1980. *The History of Wildfowling*. A & C Black. London.

Markham, Gervase. 1621. *Hunger's Prevention, or the Whole Art of Fowling by Water or Land*. London.

Markwick, W. 1795. *Aves Sussexiensis: or a Catalogue of Birds found in the County of Sussex, with Remarks*. Reprinted in 1995 from the Transactions of the Linnean Society by St Ann's Books. Malvern.

Marquiss, M. 1989. Grey Herons *Ardea cinerea* breeding in Scotland: numbers, distribution and census techniques. *Bird Study* 36: 181–191.

Martin, M. 1698. *A Late Voyage to St Kilda*. Reprinted edition. Birlinn Limited. Edinburgh.

Martin, M. 1703. *A Description of the Western Isles of Scotland*. Reprinted edition. Birlinn Limited. Edinburgh.

Matheson, C. 1963. The Pheasant in Wales. *British Birds* 56: 452–456.

Mathew, Rev. M. A. 1869. Slaughter of Sea-fowl at Weston-super-Mare. *Zoologist* 1869: 1644.

Mathew, Rev. M. A. 1877. Ornithological Notes from the West of England. *Zoologist* 1877: 177–178.

Mathew, Rev. M. A. 1894. *The Birds of Pembrokeshire and its Islands*. R.H. Porter. London.

Mathew, Rev. M. A. 1895. Pembrokeshire Birds in 1603. *Zoologist* 3rd series; Vol xix: 241–247.

Matthews, G. V. T. 1958. Wildfowl Conservation in the Netherlands. *Wildfowl* 9: 142–153.

Mayhew, H. 1861–62. *London Labour and the London Poor*. Dover publications reprint 1968. New York.

McAldowie, A.M. 1893. *The Birds of Staffordshire*. Privately published.

McCulloch, M. N., Tucker, G. M. & Baillie, S.R. 1992. The hunting of migratory birds in Europe; a ringing recovery analysis. *Ibis* 134 suppl. 1: 55–65.

Mead, W. E. 1931. *The English Mediaeval Feast*. George Allen & Unwin. London.

Melling, T. 2005. The Tadcaster rarities. *British Birds* 98: 230–237.

Melling, T. 2008. Should the Kermadec Petrel be on the British List? *British Birds* 101: 31–38.

Meltofte, H. 1993. Vadefugletraekket gennem Danmark. *Dansk Orn. For. Tidskr.* 87: 1–180 [in Danish with English summary].

Millais, J. G. 1909. *The Natural History of British Game Birds*. London.

Mitchell, F. S. 1885. *The Birds of Lancashire*. Van Voorst. London.

Montagu, G. 1833. *Ornithological Dictionary of British Birds*. A new edition edited by James Rennie. Orr & Smith. London.

Moreau, R. E. 1951. The British status of the Quail and some problems of its biology. *British Birds* 44: 257–276.

Morris, P. A. 2010. *A History of Taxidermy: art, science and bad taste*. MPM Publishing. Ascot

Muirhead, G. 1895. *The Birds of Berwickshire vol.ii*. David Douglas. Edinburgh.

Mullens, W. H. 1912. Thomas Muffett. *British Birds* 5: 262–278.

Munsche, P. B. 1981. *Gentlemen and Poachers: the English Game Laws 1671–1831*. Cambridge University Press.

Naylor, K. A. 1996. *A reference manual of rare birds in Great Britain and Ireland*. Published by the author.

Nelson, T. H. 1907. *The Birds of Yorkshire*. Brown & Sons. London, Hull and York.

Nethersole-Thompson, D. 1973. *The Dotterel*. Collins. London.

Nettleship, D. N. & Birkhead, T. R. (eds). 1985. *The Atlantic Alcidae*. Academic Press. London.

Newton, A. 1879. Hawking in Norfolk. In Lubbock, Rev. R. *Observations on the Fauna of Norfolk*. 2nd edn, edited by T. Southwell. Jarrold & Sons. Norwich.

Newton, A. 1896. *A Dictionary of Birds*. A & C Black. London.

Nicholls, S. 2009. *Paradise found: Nature in America at the time of discovery*. University of Chicago Press. Chicago.

Nicholson, E. M. 1926. *Birds in England*. Chapman & Hall. London.

Nicholson, E. M. 1929. Report on the 'British Birds' Census of Heronries 1928. *British Birds* 22: 270–323 and 334–372.

Nicholson, E. M. 1951. *Birds and Men*. Collins. London.

Nicholson, E. M. & Ferguson-Lees, I. J. 1962. The Hastings Rarities. *British Birds* 55: 299–384.

Norrevang, A. 1986. Traditions of sea bird fowling in the Faroes: an ecological basis for sustained fowling. *Ornis Scandinavica* 17: 275–281.

Oakes, C. 1953. *The Birds of Lancashire*. Oliver & Boyd. Edinburgh.

Ogilvie, M. A. 1978. *Wild Geese*. T & AD Poyser. Berkhamsted.

Parker, A. J. 1988. The Birds of Roman Britain. *Oxford Journal of Archaeology* 7: 197–226.

Parkin, D. T. & Knox, A. G. *The Status of Birds in Britain and Ireland*. Christopher Helm. London.

Pashley, H. N. 1925. *Notes on the Birds of Cley, Norfolk*. Witherby. London.

Paton, E. R. & Pike, O. G. 1929. *The Birds of Ayrshire*. Witherby. London.

Patterson, A. H. 1905. *Nature in Eastern Norfolk*. Methuen. London.

Patterson, A. H. 1929. *Wild-fowlers and Poachers: fifty years of the east coast*. Methuen. London.

Patterson, A. H. 1930. *A Norfolk Naturalist*. Methuen. London.

Payne-Gallwey, Sir R. 1882. *The Fowler in Ireland*. Field Library reprint 1985. Ashford Press. Southampton.

Payne-Gallwey, Sir R. 1886. *The Book of Decoys: Their Construction, Management and History*. Van Voorst. London.

Pearson, T. G. 1912. The White Egrets. *Bird-lore* 14: 62–69.

Penhallurick, R. D. 1969. *Birds of the Cornish Coast*. Bradford Barton. Truro.

Pennant, T. 1776. *British Zoology*. White. London.

Pennie, I. D. 1950–51. The History and

Distribution of the Capercaillie in Scotland. *The Scottish Naturalist* 62: 65–87, 157–178; 63: 4–17, 135.

Penry-Jones, J. 1958. Feathers. *PLA Monthly* 33: 91–94. Port of London Authority.

Percy, Lord William. 1951. *Three Studies in Bird Character*. Country Life. London.

Phillips, J. C. 1922–26. *A Natural History of the Ducks*. Houghton Mifflin. Boston. Dover Publications reprint 1986.

Potapov, E. & Sale, R. 2005. *The Gyrfalcon*. T & AD Poyser. London.

Potts, G. R. 1986. *The Partridge: Pesticides, Predation and Conservation*. Collins. London.

Potts, G. R. 2012. *Partridges*. Collins. London.

Pryde, P. R. 1972. *Conservation in the Soviet Union*. Cambridge University Press. Cambridge.

Pycraft, W. P. 1936. The Partridges. In Kirkman, F.B. & Hutchinson, H.G. (eds). 1936. *British Sporting Birds*. T. C. & E. C. Jack. London and Edinburgh.

Rackham, O. 1986. *The History of the Countryside*. J. M. Dent & Sons. London.

Rackham, O. 1990. *Trees and Woodland in the British Landscape*. Revised edition. J. M. Dent & Sons. London.

Ralfe, P.G. 1905. *The Birds of the Isle of Man*. David Douglas. Edinburgh.

Ratcliffe, D. A. 1973. The Dotterel as a breeding bird in England. In Nethersole-Thompson, D. 1973. *The Dotterel*. Collins. London.

Ratcliffe, D. A. 1997. *The Raven*. T & AD Poyser. London.

Raven, C. E. 1929. *Bird Haunts and Bird Behaviour*. Martin Hopkinson. London.

Ray, J. 1678. *The Ornithology of Francis Willughby of Middleton*. Facsimile edition by P.P.B. Minet, Chichely House, Newport Pagnell.

Rintoul, L. J. & Baxter, E. V. 1935. *A Vertebrate Fauna of Forth*. Oliver & Boyd. Edinburgh.

Ripley, S. D. 1965. Swans and geese. In Wetmore, A. 1965. *Water, Prey and Game Birds of North America*. National Geographic Society.

Riviere, B. B. 1930. *A History of the Birds of Norfolk*. Witherby. London.

Robinson, E. & Fitter, R. (eds). 1982. *John Clare's Birds*. Oxford University Press. Oxford

Roderick, H. & Davis, P. 2010. *Birds of Ceredigion*. Wildlife Trust, South and West Wales.

Rogers, D. 2008. The Heronries Census in Devon: 1928–2007. *Devon Birds* 61: 10–15.

Rowley, T. 1982. Mediaeval field systems. In Cantor, L. M. (ed.). *The English Mediaeval Landscape*. Croom Helm. London.

Ruffer, J. 1977. *The Big Shots; Edwardian shooting parties*. Quiller Press. London.

Saari, L. 1995. Population trends of the Dotterel *Charadrius morinellus* in Finland during the past 150 years. *Ornis Fennica* 72: 29–36.

Sale, R. & Potapov, E. 2010. *The Scramble for the Arctic: ownership, exploitation and conflict in the far north*. Frances Lincoln Ltd. London.

Salvin, F. H. & Brodrick, W. 1855. *Falconry in the British Isles*. Reprinted 1997. Beech Publishing House. Midhurst.

Saunders, H. 1899. *Manual of British Birds*. 2nd edition. Gurney & Jackson. London.

Savage, C. 1963. Wildfowling in Northern Iran. *Wildfowl* 14: 30–46.

Sclater, P. L. & Forbes, W. A. 1877. On the Nesting of the Spoonbill in Holland. *Ibis* 19: 412–416.

Scott, W. E. D. 1887. The present condition of some of the bird rookeries of the Gulf coast of Florida. *Auk* 4 (2): 135–144, 213–222, 273–284.

Scott, P. & Fisher, J. 1953. *A Thousand Geese*. Collins. London.

Seebohm, H. 1901. *The Birds of Siberia*.

Selby, P. J. 1833. *Illustrations of British Ornithology*. W. H. Lizars. Edinburgh.

Serjeantson, D. 1988. Archaeological and ethnographic evidence for seabird exploitation in Scotland. *Archaeozoologia* 2: 209–224.

Serjeantson, D. 2001. The Great Auk and the Gannet: a Prehistoric Perspective on the Extinction of the Great Auk. *International Journal of Osteoarchaeology* 11: 43–55.

Serjeantson, D. 2006. Birds: Food and a mark of Status. In Woolgar, C. M., Serjeantson, D. & Waldron, A. (eds). *Food in Mediaeval England*. Oxford University Press.

Shaw, F. J. 1980. *The Northern and Western Islands of Scotland: their Economy and Society in the 17th Century*. John Donald. Edinburgh.

Shrubb, M. 1979. *The Birds of Sussex: their present status*. Phillimore. Chichester.

Shrubb, M. 2003. *Birds, Scythes and Combines: a history of birds and agricultural change*. Cambridge University Press.

Shrubb, M. 2007. *The Lapwing*. T & AD Poyser. London.

Shrubb, M. 2011. Some thoughts on the historical status of the Great Bustard in Britain. *British Birds* 104: 180–191.

Sibley, D.A. 2000. *The Sibley Guide to Birds*. Knopf. New York.

Sim, G. 1903. *The Vertebrate Fauna of Dee*. D. Wyllie & Son. Aberdeen.

Simon, A. L. 1952. *A concise Encyclopedia of Gastronomy*. Collins. London.

Smith, A. C. 1887. *The Birds of Wiltshire*. R. H. Porter. London.

Smith, A. E. & Cornwallis, R. K. 1955. *The Birds of Lincolnshire*. Lincolnshire Naturalists Union.

Smith, T. 1930–8. The Birds of Staffordshire. *Transactions of the North Staffordshire Field Club* vols.64–72.

Smout, T. C. 2000. *Nature Contested: Environmental History in Scotland and Northern England since 1600*. Edinburgh University Press.

Somerset Ornithological Society. 1988. *Birds of Somerset*. Alan Sutton Publishing. Gloucester.

Southwell, T. 1870–71. On the ornithological archaeology of Norfolk. *Transactions of the Norfolk and Norwich Naturalists Society* 1: 14–23.

Southwell, T. 1901. On the breeding of the Crane in East Anglia. *Trans. Norfolk & Norwich Naturalists Society* 7: 160–170.

Southwell, T. 1904. On some early Dutch and English decoys. *Transactions of the Norfolk and Norwich Naturalists Society* 7: 606–617.

St John, C.W.G. 1848. *A Sportsman and Naturalist's Tour in Sutherlandshire*. John Murray. London.

Standley, P., Bucknell, N.J., Swash, A. & Collins, I. D. 1996. *Birds of Berkshire*. Berkshire Atlas Group. Reading.

Steele-Elliott, J. 1936. Local Duck Decoys. *Bedfordshire Historical Record Society: Survey of Ancient Buildings vol.III.*

Sterland, W.J. 1869. *The Birds of Sherwood Forest*. Reeve & Co. London.

Stevenson, H. 1866 & 1870 & Stevenson, H. & Southwell, T. 1890. *The Birds of Norfolk* (3 vols.). Gurney & Jackson. London.

Stevenson, H. 1871–72. Scoulton Gullery. *Transactions of the Norfolk and Norwich Naturalists Society* 1: 22–30.

Stone, D. J. 2006. The consumption and supply of Birds in late Mediaeval England. In Woolgar, C. M., Serjeantson, D. & Waldron, A. (eds). *Food in Mediaeval England*. Oxford University Press.

Stubbs, F. J. 1910a. A lost British Bird. *Zoologist* 14(4): 150–156.

Stubbs, F. J. 1910b. The Egret In Britain. *Zoologist* 14: 380–383.

Stubbs, F. J. 1913. Asiatic birds in Leadenhall Market. *Zoologist* 17(4):156–157.

Studebaker, M. 2012. Spectacled Eiders at Barrow Alaska. *Birding World* 25(2): 84–88.

Swaine, C. M. 1982. *The Birds of Gloucestershire*. Alan Sutton Publishing. Gloucester.

Tait, W. C. 1924. *The Birds of Portugal*. Witherby. London.

Tapper, S. 1992. *Game Heritage*. The Game Conservancy. Fordingbridge.

Taylor, M., Seago, M., Allard, P. & Dorling, D. 1999. *The Birds of Norfolk*. Pica Press. Robertsbridge.

Terrasse, J-F. 1964. The status of Birds of Prey in France in 1964. In: *Report of Working Conference of Birds of Prey and Owls: Caen 10–12 April 1964*. International Council for Bird Preservation.

Thirsk, J. 1997. *Alternative agriculture: a History from the Black Death to the present day*. Oxford University Press.

Thirsk, J. 2007. *Food in Early Modern England: phases, fads and fashions 1500–1760*. Hambledon Continuum.

Thom, V. M. 1986. *Birds in Scotland*. T & AD Poyser. Calton.

Thompson, D. A. W. 1936. *A Glossary of Greek Birds*. Oxford University Press.

Ticehurst, C. B. 1932. *The Birds of Suffolk*. Gurney & Jackson. London.

Ticehurst, N. F. 1909. *A History of the Birds of Kent*. Witherby. London.

Ticehurst, N. F. 1922. On former breeding-places of the Oystercatcher and Black-headed Gull in East Sussex. *British Birds* 15: 6–9.

Ticehurst, N. F. 1923. Some British Birds in the fourteenth century. *British Birds* 17: 29–35.

Ticehurst, N. F. 1934. The Swan marks of Lincolnshire. *Transactions of Lincolnshire Architectural and Archaeological Society* 1934: 59–141.

Ticehurst, N. F. 1957. *The Mute Swan in England*. Cleaver-Hume. London.

Tomialojc, L. 1997. Song Thrush. In Hagemeijer, W. J. M. & Blair, M. J. (eds). 1997. *The EBCC Atlas of European Breeding Birds: their distribution and abundance*. T & AD Poyser. London.

Toynbee, J. M. C. 1973. *Animals in Roman Life and Art*. 1996 edition published by Thames & Hudson. London.

Tyler, M. 2010. *The Birds of Devon*. Devon Birdwatching and Preservation Society.

Trodd, P. & Kramer, D. 1991. *Birds of Bedfordshire*. Castlemead. Welwyn Garden City.

Trolliet, B. 2003. Elements for a Lapwing (*Vanellus vanellus*) management plan. *Game and Wildlife Science* 20: 93–144.

Tubbs, C. R. 1968. *The New Forest: An ecological history*. David & Charles. Newton Abbot

Tubbs, C. R. 1985. *The Decline and Present Status of the English Lowland Heaths and their Vertebrates. Focus No.11*. Nature Conservancy Council. Peterborough.

Tubbs, C. R. 1992. The diaries of William Mudge, wildfowler. *Wader Study Group Bulletin* 65: 46–54.

Tucker, G. M., Heath, M. F., Tomialojc, L. & Grimmett, R. F. A. 1994. *Birds in Europe: their conservation status*. Birdlife Conservation Series no. 3. Birdlife International. Cambridge.

Turner, E. 1862. Ashdown Forest, or as it was sometimes called Lancaster Great Park. *Sussex Archaeological Collections* 14: 35–64.

Turner, E. L. 1924. *Broadland Birds*. Country Life. London.

Uspenski, S. M. 1958. *The Bird Bazaars of Novaya Zemlya*. Translations of Russian Game Reports vol.4. The Queen's Printer. Ottawa.

Ussher, R. J. & Warren, R. 1900. *The Birds of Ireland*. Gurney & Jackson. London.

Vandervell, A. & Coles, C. 1980. *Game & the English Landscape*. Debrett. London.

Van de Wall, J. W. M. 2004. *The Loo Falconry: the Royal Loo Hawking Club 1839–1855*. Hancock House Publishers. British Columbia and Washington.

Verner, Col. W. 1914. Prehistoric Man in Southern Spain – III. *Country Life July 28th 1914.*

Vesey-Fitzgerald, B. 1946. *British Game*. Collins. London.

Vesey-Fitzgerald, B. 1969. *Vanishing Wildlife of Britain*. MacGibbon & Key. London.

Voisin, C. 1991. *The Herons of Europe*. T & AD Poyser. London.

Voous, K. H. 1960. *Atlas of European Birds*. Nelson. London.

Walpole-Bond, J. 1938. *A History of Sussex Birds*. Witherby. London.

Webb, N. 1986. *Heathlands*. Collins. London.

Wentworth-Day, J. 1949. *Coastal Adventure*. Harrap. London.

Wentworth-Day, J. 1954. *A History of the Fens*. George Harrap. London.

Wheatley, J. J. 2007. *The Birds of Surrey*. Surrey Bird Club.

Whitaker, J. 1907. *Notes on the Birds of Nottinghamshire*. Walter Black & Co. Nottingham.

Whitaker, J. 1918. *British Duck Decoys of today, 1918*. Burlington Publishing Co. Ltd. London.

White, G. 1789. *The Natural History of Selborne*. (Everyman' Library Edition 1949, J. M. Dent & Sons. London).

Whitfield, D. P. 2002. Status of breeding Dotterel *Charadrius morinellus* in Britain in 1999. *Bird Study* 49: 237–249.

Williamson, K. 1945. The Economic Importance of Sea-fowl in the Faroe Islands. *Ibis* 87: 249–269.

Williamson, K. 1948. *The Atlantic Islands: the Faroe life and scene*. Collins. London.

Williamson, K. & Boyd, J. M. 1961. *St Kilda Summer*. The Country Book Club. London.

Wilson, J. 1840. The Rod and the Gun. *Tait's Edinburgh Magazine* August 1840: 477–490.

Wood, S. 2007. *The Birds of Essex*. Christopher Helm. London.

Woolgar, C. M. 1995. Diet and consumption in gentry and noble households: a case study from around the Wash. In Archer, R. E. & Walker, S. (eds). 1995. *Rulers and Ruled in late Mediaeval England*. Hambledon Press. London.

Woolgar, C. M. 1999. *The Great Household in late Mediaeval England*. Yale University Press. London

Woolgar, C. M. 2006. *Household accounts from Mediaeval England* in 2 vols. Oxford University Press for the British Academy.

Wright, C. A. 1864. List of Birds observed in the Islands of Malta and Gozo. *Ibis* 1864: 42–72.

Wright, M. 1986. Concise history of Suffolk Heronries. *Suffolk Ornithologists Group Bulletin* 72: 6–14.

Yalden, D. W. 1999. *The History of British Mammals*. T & AD Poyser. London

Yalden, D. W. & Albarella, U. 2009. *The History of British Birds*. Oxford University Press.

Yapp, W. B. 1981. *Birds in Mediaeval Manuscripts*. The British Library. London.

Yapp, W. B. 1982. Birds in captivity in the Middle Ages. *Archives of Natural History* 10: 479–500.

Yapp, W. B. 1983. Gamebirds in medieval England. *Ibis* 125: 218–221.

Yarrell, W. 1845. *A History of British Birds*. 2nd edition. Van Voorst. London.

Yarrell, W. 1876–1885. *A History of British Birds*. 4th edition, vols I & II (1876–82) edited and revised by Alfred Newton, vols III & IV (1882–85) edited and revised by Howard Saunders. Van Voorst. London.

Yeatman, L. 1976. *Atlas des Oiseaux Nicheurs de France*. Societe Ornithologique de France. Paris.

Index

Fowling with throwing sticks, from the tomb of Neb Amon at Thebes. Note the tame cat retrieving catching birds, and the egrets held as decoys (Trustees of the British Museum).

Fowling from the tomb of Nakht at Thebes, showing how the throwing sticks were used (Trustees of the British Museum).

▲ Heron-hawking with the Royal Loo Hawking Club. Here the heron is being gathered up; the bird would be examined, marked (see Figure 4.3), and then released. Note the falconers are in 16th-century dress. From a painting by Pierre Dubourcq (courtesy of Theo Daatselaar, Antiquairs).

▼ A white Gyr Falcon *Falco rusticolus*. Gyr Falcons were particularly prized for flying at Common Crane *Grus grus*; white morphs were the most expensive of all falcons.

▲ A female Eider *Somateria mollissima* incubating. Eiders take readily to artificial nest sites, which are often built for them.

◄ Freshly gathered eider down.

◄ Whooper Swans *Cygnus cygnus* on ice at the wintering grounds.

▲ A carriage gun being used to stalk Great Bustards *Otis tarda* in Germany. Such vehicles were also used hunt Cranes *Grus grus*. From Wolfgang Birkner's *Das Jüngere Jagdbuch* (courtesy of Forschungsbiblioth Gotha).

▼ Shooting partridges over pointers with flintlocks. An early 19th-century print of a painting by Sam Howitt.

Stack an Armin, St Kilda. This formidable rock was climbed annually by the men of St Kilda in rsuit of Gannet *Morus bassana* eggs and young (photo courtesy of Tony Marr).

◄ ▲ Common Guillemots *Uria aalge* (left) and Gannets *Morus bassana* were important food species all around the North Atlantic.

▲ German peasants trapping birds using a pole-net (left), clap-nets (centre), lime sticks (centre right), a flight-net (top right) and a cage-trap (bottom right). Five owls are shown as decoys, and tethered and caged call-birds are visible around the clap-nets. From Wolfgang Birkner's *Jüngere Jagdbuch* (courtesy Forschungsbibliothek Gotha).

◀ The Ortolan Bunting *Emberiza hortulana* – long prized as a table bird, and still eaten in France.

'Morning Gossip', illustrating ⊧ barber's taxidermy and some his tools of that trade. From painting by Dendy Sadler urtesy of the Thomas Ross llections).

▶ Summer-plumaged Spotted ⊦lshank *Tringa erythropus* ⊦ Red Knot *Calidris canutus*, ⊦urite specimens for 19th-tury collectors.

▲ A case of male Ruffs *Philomachus pugnax* summer plumage, mounted by the Norw[i] taxidermist T. E. Gunn. These were much soug[ht] after by collectors, due to the great variation [in] their plumage (courtesy of P. A. Morris).

◀ ▼ Little Egret *Egretta garzetta* and Kittiwake *Rissa tridactyla*, two species that were severely damaged by the plumage trade in the 19th centu[ry].